P9-BYE-267

The Silicon Valley of Dreams

CRITICAL AMERICA

General Editors: Richard Delgado and Jean Stefancic

White by Law: The Legal Construction of Race
Ian F. Haney López

Cultivating Intelligence: Power, Law, and the Politics of Teaching
Louise Harmon and Deborah W. Post

Privilege Revealed: How Invisible Preference Undermines America
Stephanie M. Wildman with Margalynne Armstrong, Adrienne D. Davis, and Trina Grillo

Does the Law Morally Bind the Poor? or What Good's the Constitution When You Can't Afford a Loaf of Bread?
R. George Wright

Hybrid: Bisexuals, Multiracials, and Other Misfits under American Law
Ruth Colker

Critical Race Feminism: A Reader
Edited by Adrien Katherine Wing

Immigrants Out! New Nativism and the Anti-Immigrant Impulse in the United States
Edited by Juan F. Perea

Taxing America
Edited by Karen B. Brown and Mary Louise Fellows

Notes of a Racial Caste Baby: Color Blindness and the End of Affirmative Action
Bryan K. Fair

Please Don't Wish Me a Merry Christmas: A Critical History of the Separation of Church and State
Stephen M. Feldman

To Be an American: Cultural Pluralism and the Rhetoric of Assimilation
Bill Ong Hing

Negrophobia and Reasonable Racism: The Hidden Costs of Being Black in America
Jody David Armour

Black and Brown in America: The Case for Cooperation
Bill Piatt

Black Rage Confronts the Law
Paul Harris

Selling Words: Free Speech in a Commercial Culture
R. George Wright

The Color of Crime: Racial Hoaxes, White Fear, Black Protectionism, Police Harassment, and Other Macroaggressions
Katheryn K. Russell

The Smart Culture: Society, Intelligence, and Law
Robert L. Hayman, Jr.

Was Blind, But Now I See: White Race Consciousness and the Law
Barbara J. Flagg

The Gender Line: Men, Women, and the Law
Nancy Levit

Heretics in the Temple: Americans Who Reject the Nation's Legal Faith
David Ray Papke

The Empire Strikes Back: Outsiders and the Struggle over Legal Education
Arthur Austin

Interracial Justice: Conflict and Reconciliation in Post–Civil Rights America
Eric K. Yamamoto

Black Men on Race, Gender, and Sexuality: A Critical Reader
Edited by Devon Carbado

When Sorry Isn't Enough: The Controversy over Apologies and Reparations for Human Injustice
Edited by Roy L. Brooks

Disoriented: Asian Americans, Law, and the Nation State
Robert S. Chang

Rape and the Culture of the Courtroom
Andrew E. Taslitz

The Passions of Law
Edited by Susan A. Bandes

Global Critical Race Feminism: An International Reader
Edited by Adrien Katherine Wing

Law and Religion: Critical Essays
Edited by Stephen M. Feldman

Changing Race: Latinos, the Census, and the History of Ethnicity
Clara E. Rodríguez

From the Ground Up: Environmental Racism and the Rise of the Environmental Justice Movement
Luke Cole and Sheila Foster

Nothing but the Truth: Why Trial Lawyers Don't, Can't, and Shouldn't Have to Tell the Whole Truth
Steven Lubet

Critical Race Theory: An Introduction
Richard Delgado and Jean Stefancic

Playing It Safe: How the Supreme Court Sidesteps Hard Cases
Lisa A. Kloppenberg

Why Lawsuits Are Good for America: Disciplined Democracy, Big Business, and the Common Law
Carl T. Bogus

How the Left Can Win Arguments and Influence People: A Tactical Manual for Pragmatic Progressives
John K. Wilson

Aftermath: The Clinton Impeachment and the Presidency in the Age of Political Spectacle
Edited by Leonard V. Kaplan and Beverly I. Moran

Getting over Equality: A Critical Diagnosis of Religious Freedom in America
Steven D. Smith

Critical Race Narratives: A Study of Race, Rhetoric, and Injury
Carl Gutiérrez-Jones

Social Scientists for Social Justice: Making the Case against Segregation
John P. Jackson, Jr.

Victims in the War on Crime: The Use and Abuse of Victims' Rights
Markus Dirk Dubber

Original Sin: Clarence Thomas and the Failure of the Constitutional Conservatives
Samuel A. Marcosson

Policing Hatred: Law Enforcement, Civil Rights, and Hate Crime
Jeannine Bell

Destructive Messages: How Hate Speech Paves the Way for Harmful Social Movements
Alexander Tsesis

Moral Imnperialism: A Critical Anthology
Edited by Berta Esperanza Hernández-Truyol

The Silicon Valley of Dreams: Environmental Injustice, Immigrant Workers, and the High-Tech Global Economy
David Naguib Pellow and Lisa Sun-Hee Park

The Silicon Valley of Dreams

Environmental Injustice, Immigrant Workers, and the High-Tech Global Economy

David Naguib Pellow *and*
Lisa Sun-Hee Park

NEW YORK UNIVERSITY PRESS
New York and London

NEW YORK UNIVERSITY PRESS
New York and London

Library of Congress Cataloging-in-Publication Data
Pellow, David N., 1969–
The Silicon Valley of dreams : environmental injustice,
immigrant workers, and the high-tech global economy /
David N. Pellow and Lisa Sun-Hee Park.
p. cm. — (Critical America)
Includes bibliographical references.
ISBN 0-8147-6709-5 (cloth : alk. paper) —
ISBN 0-8147-6710-9 (pbk. : alk. paper)
1. High technology industries—Environmental aspects—
California—Santa Clara Valley (Santa Clara County)
2. Agriculture—Environmental aspects—California—
Santa Clara Valley (Santa Clara County)
3. Alien labor—California—Santa Clara Valley (Santa
Clara County) 4. Minorities—California—Santa Clara
Valley (Santa Clara County) 5. Environmental justice—
California—Santa Clara Valley (Santa Clara County)
I. Park, Lisa Sun-Hee. II. Title. III. Series.
HC107.C22S376 2003
330.9794'73—dc21 2002013678

New York University Press books are printed on acid-free paper,
and their binding materials are chosen for strength and durability.

Manufactured in the United States of America
10 9 8 7 6 5 4 3 2 1

Contents

Preface

While typically lauded as the engine of the high-tech global economy and a generator of wealth for millions, Silicon Valley is also home to some of the most toxic industries in the nation, and perhaps the world. Next to the nuclear industry, the production of electronics and computer components contaminates the air, land, water, and human bodies with a nearly unrivaled intensity.

The Valley is also a site of extreme social inequality. It is home to more millionaires per capita than anywhere else in the United States, yet the area has also experienced some of the greatest declines in wages for working-class residents of any city in the nation. Homes are bought and sold for millions of dollars each day, yet thousands of fully employed residents live in homeless shelters in San Jose, the self-proclaimed "Capitol of Silicon Valley." Silicon Valley also leads the nation in the number of temporary workers per capita and in workforce gender inequities. Moreover, the region has an entirely non-unionized workforce and is as racially segregated as most big urban centers.

A History of Environmental Injustices

The combination of environmental degradation and social inequalities produces *environmental inequalities* and *environmental racism.*[1] Environmental racism is "the unequal protection against toxic and hazardous waste exposure and the systematic exclusion of people of color from decisions affecting their communities."[2] Environmental inequality or environmental injustice is a broader concept that includes the disproportionate exposure to hazards any marginal group confronts. From a community-based perspective, environmental injustices are policies and practices characterized by unfair treatment, discrimination, and oppression.

Social inequality and environmental disruption in Silicon Valley are intimately linked. To grasp fully the nature of environmental inequalities in Silicon Valley, one must also search for the *origins* of this system of exploitation. Accordingly, we begin by documenting that immigrants and people of color are currently concentrated in the most hazardous occupations and the most environmentally polluted neighborhoods in Silicon Valley (i.e., environmental racism). We then trace these environmental and social inequalities back across three previous historical periods (the genocide and ecocide of the Spanish Conquest of Bay Area Native nations; the racial and labor conflicts and environmental destruction of the Gold Rush era; and the labor unrest and water depletion of the agricultural period) to demonstrate that each era's political and economic structures included the subjugation of communities and workplaces populated by people of color, immigrants, and women, and the depletion of natural resources.[3]

These chapters are followed by an in-depth analysis of the present-day struggles for occupational health and safety and environmental justice in what is now called Silicon Valley. These conflicts reveal the links between social and environmental inequalities in that region's workplaces and residential communities. Our analysis is based on years of data collection, including archival research, ethnographic participant observation in a high-tech firm, and interviews with workers and labor/environmental advocates. We focus on the full range of human health and environmental impacts affecting production workers in high-tech electronics firms and related support industries. Resistance by workers and by environmental movement organizations plays a central role in this story as well. The exploitation of land and people in the region lies at the heart of the story and at the core of virtually all struggles over environmental injustice in the United States.

Because Silicon Valley has often been rocked by strife between labor and capital, communities and corporations, the powerful and the exploited, the polluters and the polluted, the region offers an unparalleled opportunity for the study of environmental justice conflicts. We document the development of these conflicts and argue that the Valley's future could be transformed by adopting promising models of social and environmental sustainability that are deeply embedded within Santa Clara County's past and its contemporary political terrain. Therefore we offer a hopeful vision of the Valley's future.

We would like to express our gratitude to the many people and organizations whose support and inspiration made this project possible. Thanks to: Larry Boehm, Leslie Byster, Carolyn Capps, Flora Chu, Dan Dohan, Carroll Estes, Frank Flores, Robert Gottlieb, Michiko Hase, Mandy Hawes, Sabrina Hodges, Evelyn Hu-DeHart, Raj Jayadev, Susan Johnson, Patricia Limerick, Jason Lum, Romi Manan, Jose Martinez, Michael Meuser, Ofelia Miramontes, Karen Moreira, Clay Morgan, Ann Morrison Piehl, Martin Sanchez-Jankowski, Raquel Sancho, Allan Schnaiberg, Rachel Silvey, Jaime Smith, Ted Smith, David Sonnenfeld, Andrew Szasz, Betty Szudy, Anna Vayr, Linda White, and our editors at New York University Press: Jennifer Hammer, Richard Delgado, and Jean Stefancic. Special thanks to Paula Levitch, our child's teacher, without whom this book would be no more than an idea. Thanks also to: Asian Immigrant Women's Advocates (AIWA), the Asian Pacific Family Resource Center in San Jose, the Labor Occupational Health Program at the University of California at Berkeley, the Robert Wood Johnson Foundation Scholars in Health Policy Research Program (University of California at Berkeley), the San Jose Public Library (specifically the California Room), the Santa Clara Center on Occupational Safety and Health (SCCOSH), and the Silicon Valley Toxics Coalition (SVTC). And, at the University of Colorado: the Council on Research and Creative Work (CRCW); the Departments of Ethnic Studies, Women's Studies, and Sociology; the Implementation of Multicultural Perspectives and Approaches in Research and Teaching (IMPART) Award Program; and the Undergraduate Research Opportunity Program (UROP). An extra special thanks goes to the dozens of persons we interviewed for this project. We dedicate this book to you and to all of the people who have struggled for their health and dignity and for environmental justice in Silicon Valley.

It is our hope that this book will help to forge links between socially, politically, and economically marginal communities and the majority of U.S. residents and workers so that we can work together as a society toward the goal of environmental justice for all.

1 Introduction

In the public consciousness, high tech is the antithesis of that old-fashioned, fossil fuel–driven industry. The news media normally discuss the new technologies as digitally clean, trafficking in information rather than goods, thriving on creativity rather than muscle. But that's a mirage.

—Christopher Cook and Clay Thompson, "Silicon Hell"

We begin with two images. The first is of a place that has been variously referred to as "The Valley of Dreams," "The Valley of the Heart's Delight," "The Garden of America," "The Garden City," "The Garden of the World," and "The Fruit Bowl of America."[1] Located mainly in Santa Clara County, California, Silicon Valley is widely hailed as the epicenter of the Information Age, the birthplace of the Digital Age, the foundation of the "new high-tech economy," and a place where people are creating, living, and enjoying the American Dream.[2] Vast fortunes are accumulated every business day in a region political leaders proudly boast is home to the largest proportion of millionaires anywhere in the world. Legends abound of "[T]hirty-year-old tycoons in T-shirts, making their first hundred million before they buy their first pinstripe suit; secretaries worth millions thanks to a few dollars spent on stock options; garage inventors suddenly finding themselves on lists of the world's richest men."[3] Immigrants from all over the world have traveled to the Valley where some have achieved unprecedented wealth by working in the high-tech industry. They have come to "experience the magic of this new engine powering the global village,"[4] the electronics sector, the world's largest and fastest growing industry. Silicon Valley is where "start-up wishes and IPO (initial public offerings) dreams set imaginations on fire. . . . There are waiting lists for luxury cars and bidding wars for palatial estates."[5] Its economy California's new Gold Rush, Silicon

Valley is a place where entrepreneurs celebrate "the days of miracle and wonder . . . [where] good times have arrived in abundance."[6] Like the first Gold Rush, all one needs in Silicon Valley is passion, commitment to hard work, and a little luck and imagination. "What a wildly democratic notion. On the Internet, anyone has a shot at success."[7] David Kratz, director of a public relations firm that does much of its business with the Valley's Internet companies, explains that in Silicon Valley, "For great ideas, there are no barriers to entry."[8] Silicon Valley is an equal-opportunity, "idea-based economy" whose principal fuel is the human imagination and ingenuity. And the payoff for many players is handsome. In 1998, the average wage in the Valley was around $50,000, while the national average was only around $31,000.[9] Wage rates were about 60 percent above—and growing far more quickly than—the national average. Homes regularly sell for $1 million more than their multimillion-dollar asking prices, and, in 1999, the Menlo Park Presbyterian Church received more than 10 percent of its contributions in stock instead of money.[10] Silicon Valley has been called "the Florence of the Information Age,"[11] where one can find "the best self-image of America."[12] Birthplace of the computer chip and the first commercially successful personal computer, "the Valley leaped fully grown into public consciousness as the answer to America's problems of a shrinking industrial base and growing unemployment."[13]

But the Valley has an underside. A second image we will consider is of a place of considerable human suffering, preventable illness and premature death, the exploitation of thousands of workers, widespread ecological devastation, and increasing social inequality. Some workers labor in their homes using toxic chemicals, earning no more than a penny for each component they attach to electronics circuit boards that power the global economy. More and more employees are temporary, which means they earn less and less each year, have little say in how they do their jobs, and can be legally fired or replaced at a moment's notice. Other workers are "downsized" and banned from the industry when they speak out about these deplorable conditions. If workers try to organize a union, they can find themselves humiliated, denounced, and fired as companies can neutralize opposition, downsize the workforce, or relocate to another state or country. Residents must be careful about the water they drink, the air they breathe, and the land they live on because all three are highly contaminated. Birth defects, cancer, respiratory ailments, and unexplained fatal illnesses are rife among workers and residents throughout the area. Those workers who own cars can look forward to numbingly long commutes along heavily polluted and congested interstates, to and from neighborhoods in which the housing costs are among the highest anywhere in the world.

Both of the above images accurately describe the same geographic space—Silicon Valley or Santa Clara Valley, California. The stark contrast between such enormous wealth, scientific innovation, and prosperity on one hand, and the relentless attacks on public and environmental health, the oppression and immiseration of thousands of workers and residents on the other, may be difficult for many of us to imagine or accept. But this is the reality of Silicon Valley. If we fail to look behind the "Silicon Curtain,"[14] however, we only see the sheen, the sleek outer shell—an image created for mass consumption by public relations firms and the mainstream media.

These two sides of Silicon Valley are also a sobering illustration of environmental racism and environmental injustice, and of the many problems that are increasingly linked to the continued globalization of the world's economies. Power, privilege, and wealth are relational, which often means that one person's riches and leisure time are derived from another's impoverishment and hard labor; one socioeconomic or racial/ethnic group's access to safe, high-salary jobs and clean neighborhoods is frequently linked to another group's relegation to dangerous, low-wage occupations and environmentally contaminated communities.[15] This is the essence of environmental racism and environmental injustice: ecological policies and practices characterized by unfair treatment, discrimination, and oppression.

In this book we take a close look at the high-technology region of Silicon Valley, California, to understand how and why this economic juggernaut came into existence and to weigh its benefits against the social and environmental costs we pay as a result of investing in this major organ of the global economy. We conclude that, while high-tech production has provided many of us with unparalleled wealth and convenience, these benefits do not justify the extraordinary exploitation of human and natural resources that make them possible. Scholars, policy makers, and activists concerned with environmental racism have thus far paid inadequate attention to immigrants, women, toxic workplaces, and the consequences of the increasing transnational mobility of workers, pollution, and firms. We enrich the existing literature by providing a new understanding of the above dynamics, drawing from research on environmental racism and inequality, immigrant labor markets, and the social and environmental nature of the high-tech revolution.

Research on Environmental Racism and Injustice

Environmental racism and injustice, which systematically exclude poor persons, immigrants, people of color, and women from decisions affecting their communities, are scourges that have burdened people around the globe for centuries. Unequal protection against toxics is what many activists are fighting *against*. But what are they fighting *for*? Environmental justice (EJ) is achieved when people are living in socially just and ecologically sustainable communities. Such communities are characterized by "decent paying safe jobs; quality schools and recreation; decent housing and adequate health care; democratic decision-making and personal empowerment . . . where both cultural and biological diversity are respected and highly revered and where distributive justice prevails."[16]

So while the terms *environmental racism* and *environmental injustice* denote the disproportionate impact of environmental hazards on marginalized communities, *environmental justice* intends to improve the overall quality of life for those same populations. For at least three decades, activists have mobilized in hundreds of communities across the United States and around the globe to document and challenge these inequalities. This mobilization is generally referred to as the environmental justice (EJ) movement.

Since the early 1970s, a growing number of scholars, activists, and policy makers have become concerned with the distributive impacts of environmental pollution on different social classes and racial/ethnic groups.[17] Hundreds of studies established a general pattern whereby environmental hazards are located in such a way that the poor and "people of color bear the brunt of the nation's pollution problem."[18] Specific findings included a strong correlation between the location of toxic facilities and communities of color in all regions of the United States and consistent lack of enforcement by the USEPA against polluters in these same communities.[19] Scientific findings such as these provided a catalyst for the EJ movement. Environmental racism became a common protest theme at the same time environmental justice (EJ) became a rallying vision.

Most EJ research has focused on the distribution of hazardous facilities in vulnerable communities and local responses to those practices (the

EJ movement). Researchers are only now beginning to explore other areas of environmental justice concerns, including the workplace, housing, and transportation,[20] but the principal subject of study remains hazardous facility siting in communities of low-income people and people of color.[21]

What about the role of immigrants in EJ conflicts? Even though recent immigrants to the United States are among the country's most socially vulnerable, politically powerless, and economically exploited populations, surprisingly little research addresses the links between immigrant communities—particularly immigrant workers—and environmental justice issues.

Immigration to what is now called the United States has proceeded almost continuously for several thousands of years. People generally immigrate for economic or political reasons (i.e., in search of jobs or as refugees or asylum seekers) or for purposes of family reunification. Asian and Latino immigrants in the United States are no different. Because the workers whose stories we tell in this book come from the Philippines, South Korea, Mexico, India, Vietnam, China, and Cambodia (among others), we present historical data and contemporary stories of people who immigrated for all of these reasons. Our justification for focusing on Asian and Latino immigrant populations is simple: they constitute the vast majority of recent immigrants in the United States, California, and Silicon Valley. While explicitly restricted or denied entry to the United States at various periods throughout history, these groups began arriving in great numbers after the passage of federal immigration legislation—the Hart-Celler Act—in 1965.[22] By 1990 Asian Americans and Latinos made up approximately 12 percent of the U.S. population[23] and more than 84 percent of new immigrants.[24] This means that, in contrast to any other period in U.S. history, the majority of newcomers today are of non-European origin—they are people of color.[25]

The social science research on immigrant labor has underscored the political and economic vulnerability of working-class immigrant populations (regardless of whether they are legal residents or undocumented persons).[26] The political vulnerability of working-class immigrants stems from both recent and historic legislation that restricts immigrant access to basic services (including health care and General Assistance) and to a number of legal protections.[27] This legislation has chipped away at the legal status of millions of immigrants and has threatened their life chances. This also means that the stability and future of entire immigrant communities are threatened. One of the most extreme forms of economic vulnerability facing immigrants is the ongoing hyperexploitation of undocumented and documented persons by employers. Immigrants face daily harassment by management, routine violations of wage and hour laws, and an exile to the lower reaches of the labor market where the jobs are highly segregated by race/ethnicity, pay a low wage, are dangerous and unhealthy, and offer scant prospects for upward mobility.[28]

Of course political and economic vulnerability frequently go hand in hand. For example, when the State of California and the U.S. Congress passed legislation restricting immigrants' rights to many social services and legal protections, these populations became more vulnerable, encouraging employers to exploit them even further.[29] The same process occurred when Congress passed the 1986 Immigration Reform and Control Act, after which the risks and penalties facing undocumented immigrant workers increased, allowing firms to pay them even less.[30] This unforgiving political and economic landscape is particularly frightening for immigrant women workers, who earn less than their male counterparts yet may be the principal caretakers of children and elderly family members.

These dynamics reveal a major contradiction in U.S. policy toward working-class immigrants: the political establishment despises immigrants and immigration, even though the reality is that immigrant labor is a core component of the U.S. economy.[31] As legal advocate Julie Su puts it, "Immigrant workers . . . provide much of the base on which the U.S. economy thrives. Unfortunately, they share in little of the profits or commodities they make possible."[32] To Su's observation we would add that immigrants also share in very little of the social and political benefits that accompany citizenship, such as the right to vote or to hold political office. Another observer writes, "[t]hat legal immigrants end up with reduced social rights compared to citizens, even though they pay regular taxes, is highly problematic for a liberal democracy."[33] This is the major contradiction with regard to immigrant workers that we, as a nation, must face in the twenty-first century.

Our Contributions

Examining the literatures on environmental racism and immigrant labor, we find several issues that need to be addressed. First, neither body of scholarship seriously addresses the relationship between immigrants and environmental justice concerns. The few exceptions include studies or reports by Gottlieb, Hunter, Hurley, and Perfecto and Velasquez.[34] Perfecto and Velasquez were among the first researchers to recognize the struggle of farm workers exposed to pesticides as an environmental justice issue. Hunter's study is the first to provide contemporary empirical evidence that foreign-born and non-English-speaking populations in the United States are much more likely to live next to toxic waste sites than the average native-born resident. Hurley and Gottlieb provide insight into how these inequalities emerged in urban areas in the United States during the nineteenth and twentieth centuries. Second, little EJ research has focused on the workplace as an environmental concern, particularly for immigrant populations. And while some research has considered this question (although not in-depth),[35] the issue of workplace EJ struggles remains undertheorized. This is a major oversight because pollution and toxics are often first produced in the workplace and those who first are exposed to and resist these poisons are workers. Third, when one examines the toxic workplace, the importance of gender becomes quite clear. The global high-tech workplace is increasingly characterized by the use of chemical toxins and the presence of a disproportionate number of women workers. However, few studies consider the role of gender in EJ conflicts beyond the predominance of women in community leadership roles. Fourth, while a few scattered studies suggest links between environmental racism in the United States and EJ struggles around the globe, such research remains woefully scarce.[36] That EJ scholars and the EJ movement are largely U.S.-based should not restrict us from broadening the scope of our work. In a global economy and on one planet with a single biosphere, environmental and immigration conflicts are rarely bound by nation-state borders.

Our reasons for writing this book are to address areas of concern that have thus far received inadequate attention from scholars studying EJ conflicts in general, and to focus on the origins and consequences of environmental injustices in Silicon Valley in particular. The principal areas of concern are: (1) the impact of environmental racism on both immigrants and people of color; (2) the study of the workplace as a site of struggle against environmental racism; (3) the role of gender in environmental justice conflicts; and (4) the global nature of environmental racism—specifically, the transnational origins and impacts of people, firms, and toxins.

Issue #1: Immigrants and Environmental Quality

It is difficult to have a conversation in the United States about immigrants and the myriad environmental injustices they suffer because so much scholarly, political, and popular media attention has been focused on the alleged social, economic, and ecological impacts of immigration itself.[37] The primary fear among the U.S. citizenry is that immigrants are displacing native-born, European American persons from their jobs and overutilizing public services such as welfare or General Assistance.[38] The recent anti-immigrant backlash has featured the passage of the Immigration Reform and Control Act of 1986, California's Proposition 187, the 1996 Immigration and Welfare Reform Acts, and the proliferation of "English-only" legislation in more than half the states.[39] Two-thirds of respondents in a 1995 Gallup poll believed that immigration to the United States should be reduced, and only around 7 percent welcomed an increase.[40] And most persons polled in the United States incorrectly believe that the majority of immigrants in this nation are undocumented.[41] This movement against immigrants thrives, paradoxically, despite recognition that the United States has always been a "nation of immigrants"[42] and despite numerous studies indicating that immigrants *under*utilize public services while providing a positive net impact on the U.S. economy.[43] This recent upsurge of anti-immigrant sentiment stems from a recurring backlash in U.S. history in which ordinary citizens, labor unions, politicians, and even environmentalists have participated.[44]

In addition to the alleged fiscal drain attributed to them, immigrants are often believed to cause grave ecological damage as well. The alleged environmental cost of immigration is a charge levied by many nativist organizations, a dimension of the anti-immigrant backlash with deep historical roots.[45] In *Alien Nation,* one of the most widely cited anti-immigration treatises of the last decade, author Peter Brimelow writes, "the immigration resulting from current public policy . . . is attended by a wide and increasing range of negative consequences, from the physical environment to the political."[46] In the early 1990s, Republican congressional leaders released press statements borrowing environmentalists' views linking immigration (generally associated with "population control") to strains on the nation's "carrying capacity" and natural resource reserves. The Republican leadership even co-opted the environmentalist concept of "bioregionalism" to argue that populations should remain in place rather than migrate, so as to prevent an undue burden on the U.S. ecosystem.

But conservative politicians do not deserve all of the blame, because many environmentalists and environmental organizations in the United States support restrictionist immigration policies.[47] For example, many leaders within the Sierra Club and the Zero Population Growth organization have made it known that immigrants are unwelcome in the United States because of their threat to "our" carrying capacity. This policy position led to a near-catastrophic schism within the Sierra Club's leadership, its membership, and affiliated chapters around the nation during the mid-1990s. Dave Foreman, the founder of the environmental group Earthfirst!, declared publicly that he favored limiting immigration because of its supposed association with environmental harm. The radical environmental movement icon Edward Abbey (author of the classic *The Monkey Wrench Gang*) staked out the same position. Even more disturbing, many leaders of the contemporary eugenics movement—including the father of sociobiology, Harvard professor E. O. Wilson, and Garrett Hardin, the former vice president of the American Eugenics Society and honorary chair of the group Population-Environment Balance—have had a major influence on this debate and on the U.S. environmental agenda.[48] Contrary to these claims, the evidence suggests that, as with fiscal and public service resources, immigrants leave a much *lighter* ecological "footprint" in the United States, particularly when compared with the high rate of consumption and energy use of the average native-born person and the even greater use of resources by the military and major corporations.[49]

Furthermore, as some scholars have shown, immigrants bear the brunt of the nation's pollution in their communities and their workplaces.[50] For instance, in Silicon Valley, 70 to 80 percent of the production workforce is composed of Asian and Latino immigrants, most of whom are women.[51] Since the mid-1970s, epidemiological studies have emerged detailing alarmingly high rates of occupational illness among Silicon Valley production workers—rates more than three times that of any other basic industry.[52] These illnesses include respiratory disorders, miscarriages, birth defects, and cancer, to name only a few. While making the microchips, printed wire and circuit boards, printers, cables, and other components of the electronics hardware that millions of consumers and businesses depend upon, these workers are often exposed to upwards of seven hundred different chemicals.[53] Masking this backstage of toxic horrors is a major public relations campaign promoting Silicon Valley industries as "the clean industry" (see chapters 4, 5, and 6). This popular image of electronics as clean and pollution-free emerged as a selling point to potential host communities who had historically been inundated with pollution from traditional manufacturing sectors, such as auto and steel.

In this book we seek answers to questions about the immigrant-environment nexus, such as: What are the origins of environmental racism in Silicon Valley? Why and how do immigrants become the targets of environmental racism and inequality? In what ways are immigrant workers and residents resisting this exploitation? Can one find examples of multiracial collaborations among and between immigrants and native born groups to redefine and reclaim their workplaces and communities?

Issue #2: The Workplace and Environmental Injustice/Racism

The majority of environmental justice research considers local communities as the primary unit of analysis. But since factories and firms create enormous volumes of pollution, it is surprising that few studies focus on the workplace, or what Karl Marx called "the point of production." By the same token, an extensive literature focuses on hazardous work without addressing the environmental justice dimension. Only a small number of EJ studies examines the workplace. Important links need to be made between these bodies of research.

The growing literatures on environmental justice and occupational health demonstrate that, in addition to being subjected to disproportionately high environmental impacts in their communities, people of color, the poor, and immigrants tend to confront similar hazards at work.[54] As with EJ studies of hazardous facility siting, many studies of hazardous work and social inequality were initiated, in part, as a response to social movements. The best example is that of farm workers and pesticide poisoning. The United Farm Workers and their charismatic leader, Cesar Chavez, introduced the farm worker dilemma to the public in the 1960s. Reports later estimated that 1,000 farm workers die each year from pesticide poisoning, and more than 300,000 become ill.[55] Moreover, this population is about 90 percent people of color and/or immigrants.

While not focused on EJ issues per se, much of the research on workplace hazards and social stratification—lends support to the environmental racism thesis through the use of existing large data sets.[56] For example, using the Panel Study of Income Dynamics, the Current Population Survey, the Survey of Economic Opportunity, the Quality of Employment Survey, and the National Longitudinal Survey, studies have demonstrated that people of color—particularly African Americans and Latinos—face greater threats on the job than their white counterparts, even when controlling for education levels.[57] People of color, immigrants, and politically marginalized ethnic groups currently and historically occupy the lowest status, highest risk, and lowest wage jobs in this society. In every way, the workplace is an environmental justice issue.

Other research, while recognizing racial inequality, proposes a more class-oriented thesis of occupational health and safety, arguing that the practice of maintaining poor working conditions must be understood within a broader theory of capitalist exploitation of the working class.[58] For example, Navarro develops a class theory of health and stress under advanced capitalism and argues that the capitalist class engages in the "absolute expropriation of health from workers."[59]

We seek answers to such questions as: What are the sociological and political implications of recognizing the workplace as an environmental concern? In what ways are workplace and community-level toxic exposures linked?

Issue #3: Gender as a Factor in Environmental Justice Conflicts

The concentration of people of color in hazardous, health-compromising jobs is well documented.[60] However, the relegation of women—particularly immigrant women and women of color—to hazardous jobs is a core component of this process and deserves equal attention. Accordingly, we devote equal consideration to the role of gender in Silicon Valley's environmental justice conflicts. Throughout Santa Clara County's history, the workplace, environmental inequality/injustice, and resistance to this exploitation have displayed a striking gender dimension. EJ scholars have made solid contributions to our knowledge of gender and environmental conflicts by considering the importance of mobilizations by women, particularly mothers, against unfair working and living conditions.[61] We expand upon these earlier important studies to demonstrate that gender also matters a great deal in the way men and women are exposed to hazards *in the workplace*. Traditional environmental justice studies are strictly limited in their potential to illuminate the links between gender and toxic exposure. This is because most of these studies correlate race or class with toxic waste sites, using census data. The problem is that, while communities can be described, and neatly separated, as "poor/low-income," "working-class," or "African-American or Latino," and so forth, few "female communities" appear on any map. So, by their very design, these traditional EJ studies overlook that women are regularly targeted for—and suffer from—environmental inequalities. Yet certain "communities" feature women concentrated in close proximity to toxics: workplaces all around the world are highly gender segregated, and in many the occupants labor with chemicals day in and day out. Toxic workplaces offer precisely the type of data that traditional EJ studies are unable to access.

Gender inequality plays a major role in the distribution of power and risk in Silicon Valley industries, with white males filling the majority of managerial and ownership positions, and women—mostly immigrants of color—concentrated in the lower-paid, highest hazard occupations.[62] Most immigrant women workers in the Valley came to the United States seeking an improvement in their economic situation and lack the skills and social networks to move up and out of these jobs. In chapters 4 through 8, we present evidence that this gender segregation in the industry is the result of a lack of economic opportunity and of conscious and selective recruiting on the part of management.[63]

If these gender segregated workplaces are indeed highly toxic, we must evaluate the health consequences of these occupational hazards. Beginning in the late 1970s, studies emerged that documented the highly toxic and dangerous nature of high-tech work. Most of the studies that made headlines (and led to reform of corporate policies regarding toxics use) focused on the negative impacts of chemicals on female workers' reproductive systems (see chapter 5). Miscarriages, birth defects, sterility, distorted menstrual cycles, toxic breast milk, and breast cancer are only a handful of the illnesses and disorders women electronics workers face.

Gender also plays a major role in the way Silicon Valley corporations handle reports and protests about toxic work. Male managers frequently reassure women workers that their suspicions about job-related health problems are baseless and nothing more than "mass hysteria"—even though many of their coworkers may be ill and dying.[64] Electronics industry managers often view immigrant women and women of color as second-wage earners (and therefore justifiably paid less), socially and culturally compliant, less likely to agitate for benefits, more physically adaptable to monotonous and intricate labor tasks, and easier to control.[65] All of these stereotypes have been proven inaccurate in various studies; many high-tech women workers are single mothers (not second-wage earners) or primary wage earners even when married, are just as likely to agitate for their rights as men, and are no more physically adapted to intricate labor tasks than any other group of people.[66]

Grassroots resistance against environmental injustices in Silicon Valley would not be the same without the presence of women activists. Women have been at the forefront of organizing against the high-tech industry in communities and workplaces in Silicon Valley, the United States generally, and around the world. Organizations like Asian Immigrant Women's Advocates (AIWA), the Santa Clara Center for Occupational Safety and Health (SCCOSH), the Silicon Valley Toxics Coalition (SVTC), Working Women's Leadership Project (WeLeaP!), the GABRIELA Network, and many others were started by women or linked to women-led organizations and have taken on the electronics industry locally and globally. Moreover, the thousands of less visible labor actions on the shop floors in the Valley are frequently initiated by women workers.[67]

Many explanations may be offered for the predominance of women in community-based EJ movements. Women are generally the primary caregivers for their children and are therefore the first to notice when their daughters or sons become environmentally or chemically sensitive. Because of their lower body mass, children are much more susceptible to toxics than adults. As caregivers, women are also much more likely to be participants in community affairs and neighborhood organizations, prior to the recognition or onset of environmental contamination. So when an environmental threat endangers a community, women are often already in the best social position to take notice and to mobilize against it.[68]

Toxics in the workplace and community often appear as distinct phenomena. In this book, we constantly link them in order to illuminate the importance of both arenas as sites of struggle for environmental justice.[69] This workplace-community nexus has a gendered dimension as well. For example, the first health studies of contaminated communities in Silicon Valley (after a chemical spill at the Fairchild Semiconductor company in 1982) and subsequent health assessments of communities in close proximity to toxic Superfund sites revealed that birth defects and other reproductive disorders were elevated in both cases (see chapters 4 and 5). Hence, women and their children are consistently on the front lines of chemical exposure and environmental justice struggles in Silicon Valley.

Not only do women constitute the majority of production workers in high-tech firms; they also account for the majority of temporary workers, the majority of workers doing piecework in their homes, and the majority of workers in low-status occupations that support the electronics industry and the regional economy, such as housekeeping, janitorial, and health care jobs. Silicon Valley's electronics industry has been called "one of the last great bastions of male dominance."[70] Notwithstanding the existence of a few token female high-tech CEOs, patriarchy affects women at all levels of the industry. For example, not only are women concentrated in the most hazardous, low-wage jobs, but the Valley's industry also has an abysmal record of providing childcare and can boast of virtually no women on its corporate boards of directors.[71]

How and why do women become the targets of environmental inequalities? How does the role of gender alter existing models of environmental injustice? Any discussion of environmental injustice in the workplace must integrate the gendered components of environmental risk.

Issue #4: Transnational People, Firms, and Toxics— Global Environmental Inequalities

As we noted earlier, the research on environmental inequality/racism has made only a few inroads into transnational EJ struggles. Not surprisingly, the scholarly literature on immigrant labor has considered the transnational nature of human mobility for quite some time. Immigrants are, by definition, transnational persons. They also often have transnational identities and families, having spent significant time in more than one nation. They often conduct transnational economic transactions via remittances back home, in many cases supporting entire communities and regional economies.[72]

In what ways does this transnationality bear on environmental issues? Many of the source countries Silicon Valley immigrants come from suffer major environmental disruptions at the hands of the U.S. military and a host of transnational corporations. Much of the widespread chemical contamination of the air, land, and water in the Philippines, Mexico, India, South Korea, China, and Vietnam stem from war-related and economic development activities initiated by the U.S. military, U.S.-based transnational firms, and the World Bank and International Monetary Fund. Mexico City, the world's most populous city, has such overwhelming air pollution problems that in recent times, the mayor proposed constructing massive fans to blow away the contaminants. The hundreds of industries that have sprung up along the U.S.-Mexico border—the *maquiladoras*—are responsible for dumping tons of toxic waste into the environment in which workers must live and labor, and the volume of toxics produced and dumped has increased rapidly since the implementation of the North American Free Trade Agreement (NAFTA) in 1994.[73] Seoul, South Korea, from which more than 50 percent of the new Korean immigrants hail, "is worse than most American metropolitan cities in terms of pollution, traffic congestion, crime, and housing conditions."[74]

One Silicon Valley worker told us how his physically deformed and mentally disabled child was exposed to agricultural pesticides (many of which are produced by U.S. firms) during her first years in Vietnam, when they lived as farmers. China's wholesale displacement of rural persons via the construction of the Three Gorges Dam is funded in part by the Export-Import Bank and U.S.-based Morgan Stanley Corporation, and the absorption of thousands of poor Chinese into low-wage, toxic sweatshops in urban areas in China is facilitated by U.S. and other transnational corporations' incessant thirst for low-wage labor. The Philippines has been the site of extreme chemical contamination ever since the U.S. military set up major bases there during the twentieth century.[75] In the Free Trade Zones of Manila and other cities, thousands of young Filipino women labor long hours for little pay in highly toxic electronics and other global industries. Many of these women immigrate to the United States and work for the same or similar companies under conditions not much better than those in the Philippines. In Vietnam, nearly one-third of the population was internally displaced during the war with the United States, "and over half of the total forest area and some 10 percent of the agricultural land was partially destroyed by aerial bombardment, tractor clearing, and chemical defoliation";[76] furthermore, the short- and long-term human health and ecological consequences of dumping more than 11 million gallons of Agent Orange (produced by U.S. chemical firms) are immense. In addition to the normal trauma and stress associated with the immigration process (particularly if one is a refugee), much of Silicon Valley's immigrant population is receiving a "double dose" of toxics: at work and at home here in the United States as well as in their country of origin.[77]

Silicon Valley is a global city in which the people, money, businesses, and even the pollution have global origins and impacts. Home to many transnational corporations with offices, production facilities, subsidiaries, and investments spanning the globe, many of Silicon Valley's managers regularly travel from nation to nation on business, family visits, and vacations. Many production workers and white-collar employees in the Valley are immigrants from Asia and Latin America who regularly return to their home countries for a variety of personal and professional reasons. Some of these workers were even employed at electronics/computer companies in their home countries prior to immigration, so the trip to Silicon Valley represents a promotion to the central office rather than an entirely new experience.

What are the transnational social and environmental costs of a globalizing electronics industry? What problems arise for workers and governments when firms operate "beyond flag and country" and have no geographic base? How can workers and environmentalists address the problems of the global production and disposal of toxic waste in this industry?[78]

Why Environmental Justice? Why Silicon Valley? Why Immigrant Workers?

A number of recent books address environmental justice. Under the category of "traditional" environmental justice struggles (i.e., against landfills, incinerators, etc.), Roberts and Toffolon-Weiss (2001) and Cole and Foster (2001) are the superlative works. Both of these books make unique contributions to the study of the EJ movement and add to our knowledge of the causes of environmental injustice and the consequences of activism on community politics. Our book extends these perspectives by adding a richer historical analysis of environmental injustice and casting a broader net that encompasses the roles of gender, several immigrant communities, the workplace, and EJ conflicts outside the United States.

Camacho (1998) and Mutz, Bryner, and Kenney (2002) include a number of case studies of environmental injustices and their relationship to natural resources such as water and land. These volumes are among the first to transcend the traditional waste facility siting approach to environmental inequality with an emphasis on ecosystems. Our book builds on these important works by providing a deeper case study and ethnographic focus on a single region and ecosystem, achieving a greater balance between social justice concerns and natural resource protection.

Scores of other studies and books focus on Santa Clara County and the Silicon Valley area. Most published work about Silicon Valley consists of books proclaiming the many virtues of the high-tech industry, with its unparalleled wealth and infinite examples of business innovation.[79] Because our point of departure is the way the high-tech industry's social and environmental harm outweigh many of its economic benefits, we question the most fundamental assumptions upon which these works are based. Some studies, however, critically examine the past and present political economy of the region. These include research on labor conflicts among ethnic groups in the Valley's cannery industry during the twentieth century;[80] studies of race, class, and gender stratification and working conditions in the electronics industry;[81] and policy research on social and environmental problems associated with the growth of the electronics industry in the area.[82] However, virtually no research has viewed the Silicon Valley phenomenon as an environmental justice issue—that is, linking social conflicts in workplaces and communities with struggles over environmental quality.[83]

This book complements and builds on these previous studies in three ways. First, as the only major study of Silicon Valley and Santa Clara County from an environmental justice theoretical approach, it explores the links between the exploitation of immigrants, women, people of color, and their physical environments and the degradation of the natural ecosystem. Second, this is the only major research effort to produce an integrated environmental and social history of Santa Clara County. We reexamine the Spanish Conquest, the Gold Rush, the Agricultural period, and the emergence of the computer-electronics sector as forms of development that produced environmental inequality and environmental racism, rather than simply as chronological stages in the region's economic evolution. The links between environmental justice and history are crucial because the vast majority of EJ research overlooks history and assumes that environmental racism emerged no earlier than the 1980s. By presenting the evidence that immigrants and people of color have been battling environmental inequalities for centuries, we challenge the common wisdom concerning the very nature of environmental racism.

Third, while many researchers have noted the transnational reach of many of the Valley's electronics firms and workers, we combine those observations with the reality that the high-tech industry's pollution has global impacts as well—producing transnational forms of environmental inequalities. The implications of this last dimension of the book are profound because it underscores that while immigrants, people of color, and Third World nations have borne the brunt of environmental injustices over the last half millennium, it is becoming increasingly clear that today all are paying the consequences.

The Symbolic and Structural Importance of Silicon Valley to the Global Economy

If Silicon Valley is the core of the computer-based, Internet, and information economy—a sector generally believed to be the cutting edge of both the U.S. and the broader global economies—garnering a thorough understanding of how it works enables us to comprehend larger social, political, economic, and environmental changes in the United States and abroad. Heads of state and ministers of finance from all over the world regularly visit Silicon Valley to learn from the top high-technology firms and to recruit businesses to their home countries with the ultimate hope of replicating the Valley in one form or another.[84] As one author writes, "[a]ttracting high tech has become the only development game."[85] Silicon Valley has long been the model for postindustrial, post-smokestack, high-technology economic development, with replicas such as Silicon Hills and Silicon Prairie in Texas, Silicon Desert in Arizona, Silicon Mountain in Colorado, Silicon Glen in Greenock, Scotland, and the clusters of electronics firms springing up in Ireland, Germany, Taiwan, South Korea, and other places around the world. Therefore, our discussion of the plight of workers and the ecosystem in Silicon Valley, California, has relevance for communities from California, the Caribbean, and Latin America to Europe and Asia. In the language of social science, the conclusions and findings that we present in this study are, in many ways, generalizable to a range of other locations around the globe.

The "Valley of the Heart's Delight" is often referred to as the "Valley of the Toxic Fright" by environmentalists and occupational health advocates.[86] The Valley holds many dubious distinctions, including hosting the highest density of federally designated toxic Superfund sites anywhere in the nation and consistently ranking among the top cities in the United States with regard to urban sprawl patterns, average time spent driving on the freeway, and auto-related smog.[87]

Since the eighteenth century, Santa Clara County has also been the site of major conflicts among Native peoples, European conquerors, workers, immigrants, women, environmentalists, the state, and industry over various labor and environmental concerns. These struggles began with the Spanish conquest of Native peoples in *Alta California*, and continued through the Gold Rush, the many farmworker and cannery strikes in this once fertile agricultural region, and the battles for occupational safety and environmental justice in the late twentieth century.

Environmental inequalities in Silicon Valley's electronics industry are the result of a confluence of several factors: (1) historical patterns of labor migration and the concentration of a multi-ethnic, immigrant, and largely female workforce in the region; (2) industry's introduction of environmentally destructive technologies coupled with authoritarian control over the labor process; and (3) the degree of autonomy, organization, and resistance among workers and social movement organizations. These factors play major roles in the continuing battle over private profit, good jobs, and natural resources in Santa Clara County, a battle that began in 1769.

Since 1976, California has been the top destination for new immigrants to the United States.[88] In the year 2000, California became a "majority minority" state, with people of color constituting more than 50 percent of the population. If the United States is a "nation of immigrants," then California is the ultimate measure of the state of immigration. California is a multiracial, multinational, multilingual, multicultural society, home to one-third of all immigrants in the United States, and it boasts one of the world's largest economies. That economy is fueled in large part by natural resources and immigrant labor.[89] The white population of Silicon Valley is now less than 50 percent, while Latinos and Asian/Pacific Islanders comprise 24 percent and 23 percent, respectively.[90] Asian/Pacific Islanders represent the fastest growing demographic group in the area, and even these figures understate the high numbers of immigrants working in Silicon Valley industries who commute daily from outlying (more affordable) communities.

If the war on immigrants and the environment is a fundamental part of the broader war on *all* workers and on the global environment,[91] immigrant workers and communities are the front-line casualties of this general assault rather than its root cause. As immigrants' legal protections and working and living conditions erode, we have witnessed the same trends among nonimmigrant workers and residents in Silicon Valley and across the United States. The state and capital have worked together to ensure that both immigrants and native-born populations receive less and less in wages, have fewer and less powerful unions, and enjoy reduced legal rights. At the same time, assaults on the integrity of the ecosystem continue virtually unabated, and many of the most toxic technologies are generally exported to the Global South when the U.S. environmental movement brings pressure to bear on polluters in the states. Therefore the native-born population in the United States should be concerned about the treatment of immigrants and the environment for humanitarian as well as self-interested reasons.

Methodology and Overview of the Book

We were attracted to this project because it enabled us to pursue longstanding interests in environmental justice and occupational health as well as immigrant labor and women's health. On a more personal note, both of us come from immigrant families, and the parents of one of us actually worked in major Silicon Valley high-tech firms.

We gathered data for this study between 1998 and 2001. We drew on four principal research methods. First, we conducted a review of the literatures on environmental justice, immigrant/women's labor, and the electronics industry in Silicon Valley and other nations. Based on this review, we determined that the four major themes mentioned earlier have not been sufficiently addressed in EJ studies. Second, from several archives we conducted systematic content analyses of newspaper articles, government documents, books, and manuscripts on environmental conflicts, labor struggles, and racial strife in Santa Clara County from pre-European contact to the present. Sources for these data include: the University of California-Berkeley library system, the City of San Jose Public Library, and the Labor Occupational Health Program Library at UC Berkeley. The stories presented in this book are also based on hundreds of memos, reports, internal documents, and studies from various grassroots and advocacy organizations. One of the most important data sources was the series of legal depositions to which Amanda Hawes allowed us access.[92] These depositions contain the firsthand testimony of dozens of ill and dying high-tech workers engaged in litigation against transnational firms. Third, we gathered data through participant observation. This involved fieldwork in a Silicon Valley firm for several weeks and volunteering at—and working for—community-based EJ organizations in the area for more than three years. This type of ethnographic fieldwork presents as many challenges as opportunities, but we firmly believe that nothing can substitute for firsthand experience with a subject as sensitive and disturbing as environmental racism. Fourth, we conducted semistructured interviews with workers and labor/environmental activists directly engaged in EJ-related conflicts in Silicon Valley. We were allowed access to workers and advocates through the kindness of organizations like the Santa Clara Center for Occupational Safety and Health (SCCOSH) and the Silicon Valley Toxics Coalition (SVTC).

A growing movement of scholars is arguing that a new methodology must take root in the academy to respond to the crises and needs of communities. This methodology is *advocacy* or *participatory research*, the theory and practice of making the scholarly enterprise more application-oriented, more sustainable, and more relevant to communities.[93] Advocacy researchers argue that scholars ought to (1) begin viewing communities not only as research subjects and informants, but also as partners in the research endeavor; and (2) give something back to these communities. Scholars practicing advocacy research, therefore, view themselves, their students, their research informants, and the public as stakeholders in the knowledge production process. While we do not presume to practice pure advocacy research, this book was produced in the spirit of this tradition. Both of us were intimately engaged in working on behalf of immigrant rights and environmental justice organizations in Silicon Valley during and after the data gathering component of this study. We therefore view ourselves not only as scholars, but as activist-scholars.[94]

The remainder of the book is straightforward. Chapters 2, 3, and 4 present a social and ecological history of Santa Clara County. In these chapters we offer support for our thesis that immigrants, people of color, women, and their labor are a core feature of environmental justice struggles in the Valley. Chapters 5 through 8 offer data to illuminate our claim that environmental justice research must place the workplace environment at the center of analysis. In these chapters, the focus is on the electronics and related industries in Silicon Valley. We present data that explain how and why Silicon Valley jobs expose immigrants, women, and people of color to some of the most dangerous and toxic substances seen in any known occupation. In chapter 8, we analyze the electronics industry's success at empire building and colonization around the globe. We discover that many of the same patterns of environmental injustice occurring in Silicon Valley have spread rapidly to other nations. We contend that if environmental justice research is to advance significantly, the units of analysis must move beyond community and nation-state borders. In chapter 9 we evaluate efforts to resist this process led by workers, environmentalists, and communities. We consider the promise of the many social movement campaigns, policy reform efforts, and alternative models of development that have shaped the Valley and the industry for the better. In chapter 10, we bring together the four major themes of this study to produce a coherent statement about EJ struggles in Silicon Valley and the high-tech global economy. Throughout the book, we present evidence that workers, environmentalists, and residents of Santa Clara County and other nations where high-tech has spread have often resisted social and environmental injustices imposed by the state and industry. These data put to rest any notion that the alleged economic miracle of Silicon Valley and the Information Age is uncontested and untroubled.

2

Early History and the Struggle for Resources

Native Nations, Spain, Mexico, and the United States

> The struggle for land has always been the hallmark of politics in North America. Those who control the land control the resources.
> —Ward Churchill, *Struggle for the Land*

Introduction

In 1978, at the Signetics Corporation plant in Sunnyvale, California, several women employees informed the management of their concerns about noxious fumes on the shop floor. Signetics was a major Silicon Valley player and a large manufacturer of semiconductors for military, business, and consumer markets. Initially, the company maintained that these women workers, who experienced dizziness, nausea, mental confusion, and emotional problems, were suffering from what Signetics termed "assembly line hysteria." However, after many male employees also came forward with similar concerns, the company hired an industrial design firm to investigate the fumes. Like canaries in a coal mine, the three most significantly affected employees (all women) were shifted around the plant in an effort to detect the most serious areas of concern. They became known as the "Signetics Three." Eventually the National Institute of Occupational Safety and Health (NIOSH) conducted a study and concluded that significant occupational health problems were evident at Signetics. This finding would not surprise any physician who was aware that Signetics used scores of toxic chemicals in the production of semiconductor chips. The company later fired the "Signetics Three" and has the dubious distinction of having polluted the earth so thoroughly that the USEPA has designated one of its properties a federal toxic Superfund site.[1]

This case underscores that the region of northern California known as Silicon Valley is a site of intense ecological devastation and human exploitation. Thousands of workers and residents, mostly women, immigrants, and people of color, face exposure to hundreds of toxics on the job and in their communities every day. This is environmental injustice with a high-tech face. In this chapter we provide a historical basis for our claim that immigrants, people of color, and their labor are at the center of environmental justice conflicts in Silicon Valley. In so doing, we argue that these populations have confronted environmental injustices in the region for more than two hundred thirty years because (a) they have frequently had their land and natural resources taken from them and destroyed; (b) they have often been denied citizenship and therefore have little formal political power; and (c) they have been concentrated in enslaved, indentured, and related "free" exploitative labor markets where wages are nonexistent or very low and the risks to one's health are substantial. By presenting evidence that immigrants and people of color have been battling environmental inequalities for centuries, we challenge the common wisdom that environmental injustice is a recent phenomenon.

Environmental justice (EJ) conflicts in Silicon Valley have been a hallmark of the region since 1769, when dominant groups controlled and degraded both natural resources (such as land, minerals, and water) and the labor of indigenous, immigrant, and other populations of color, while also imposing environmental risks on these groups. By demonstrating the historical continuity of EJ conflicts in this region, we also address the question of the origins of environmental injustice in Silicon Valley. Simply put, environmental injustice in the area can be traced to the Spanish conquest and its associated devastation of Native American populations and Bay Area ecosystems.[2] Since that time, each subsequent period has built upon the previous era, and the exploitation of human labor and natural resources has remained constant.

Native American Communities

Any history of the present-day Silicon Valley region of California must begin with a consideration of First Nations. Long before the Spanish Conquest, beginning at least as early as 500 C.E., the San Francisco Bay was inhabited by an estimated fifty independent nations of the Ohlone/Costanoan people.[3] The Ohlone economy was largely based on the accumulation, consumption, and trade of acorns and shellfish. In fact, these societies used meticulously carved shells as currency.[4] Trade routes for dried and fresh shellfish, shell jewelry, Sierra obsidian, and coastal abalone shell extended throughout the San Francisco Bay Area, northward through the Sacramento and San Joaquin River valleys, and to points east.[5] The Ohlones also used burning techniques for agriculture, specifically for planting and harvesting wild grass and seeds.[6] This Native American nation did well as a result of the abundance of natural resources and experienced relatively few economic hardships prior to the Spanish arrival.[7]

As with many traditional cultures, Bay Area Native peoples had an economy that was tied to local natural resources and an ethic that militated against the destruction of this resource base.[8] The Ohlone people built their homes, sweat lodges, and ceremonial structures out of organic materials that grew in the immediate vicinity, such as tree bark, brush, and earth. Thus, the "Ohlone enjoyed an abundant life. . . [and] the subsistence strategy that they enjoyed represented a successful adaptation to a rich but expansive environment."[9] The Ohlones' subsistence environmental ethic extended beyond the use of natural resources, beyond fishing and acorn farming, and was applied to the extraction of mineral resources as well:

> The most important function of Indian mines and quarries was to supply materials for the implements and tools to be used to secure subsistence . . . or maintain life's necessities . . . or as luxuries. . . . The Indians probably had no particular "mining code" as such, although one principle of Indian philosophy which ran through everything was applied to the use of mines and quarries; this principle is *the conservation of natural resources.*[10]

Moreover, with regard to the question of public versus private access to these natural resources, in many cases there was a communal ethic guiding early Native California communities: "there was no private, individual ownership of mine or quarry sites by the Indians. Rather these places were considered property of the group, in which everyone was allowed to share."[11] This communal ethic was also the norm for food collection and distribution, as evidenced during the Ohlone's fall acorn harvest.[12]

Native Americans regularly drew on mines for materials like turquoise, obsidian, and soapstone. One mine—later called the New Almaden by the Spanish, Mexicans, and Anglos—was especially popular due to its abundant supply of the bright substance cinnabar (clay containing mercury), which was used to produce red paint for body decorations.[13] The New Almaden Mine, located in present-day Silicon Valley, later gained worldwide significance for its central role in the California Gold Rush.

Thus, Native peoples in the Bay Area sustained their societies by drawing on the region's impressive ecological wealth *and* by constructing a social system based on the equitable distribution of resources and limits on the use of that natural wealth. Natural resource conservation and a "commons" system of access to materials are two features of California Native economies that would soon be a thing of the past. On the eve of the Spanish conquest, the Ohlones were drawing their last free supplies of fresh water from the Guadalupe River near land that would later become El Pueblo de San Jose.

Native Americans had been living in what is today known as California for roughly fifteen thousand years prior to the arrival of explorers from New Spain. The indigenous population of the state during the late eighteenth century was an estimated 310,000 people who spoke nearly a hundred distinct languages.[14] Early European explorers viewed the area as a "pristine 'wilderness' shaped entirely by the hand of God"[15] and populated by "heathens" whose proper fate would be conversion to Christianity or death. They aimed to tame this wilderness and harness its abundance for the growth of a divinely ordained New World civilization.

Sacking El Dorado: The Spanish Conquest, Mexican Independence, and Native Subjugation

The first peoples in the Americas were, in many ways, "the ultimate keystone species" whose genocide altered not only these early civilizations, but entire ecosystems throughout North America as well.[16] The conquest of Native Americans went hand in hand with the domination of nature.

Spanish explorers arrived in *Alta California* (upper California) in 1769 from colonial strongholds in present-day Mexico and the southwestern United States. When they began to explore the region, they witnessed thriving native civilizations supported by a rich ecosystem. As one such explorer's journal indicated:

> There is a positive maze of very large freshwater lakes with a great deal of swamp and brush patches in this hollow; and I know not how many large running streams, and two or three very large heathen [Native] villages. . . . They had several tule-rush floats with oars, with which they fished in the lakes. . . .We came to a very full-flowing stream which it took us some trouble to cross because it had so much water, and close to this was a very big heathen village. . . .[17]

> We have come across four or five villages of very fine well-behaved heathens . . . all of whom were dying for us to come to their villages and have them feed us. We have passed through three or four of the villages and in each one they have presented us with a great many servings of [roots and nuts].[18]

Despite the racist and demeaning use of the term "heathen," some of these Spanish explorers of the area were obviously impressed with both the region's bountiful natural resources and the generosity of the Ohlone people. Both would be taken full advantage of soon thereafter.

The Spanish Crown charged its representatives with the task of developing a network of missions, military presidios, and settlements that could ensure Spain's control over the entirety of California. The major military threats to this expansion were from the British and Russian settlements just north of the region, so the Spaniards aimed to be the first to claim the area. In other words, this expansion of territory north of Mexico was colonization and empire building in progress—an early exercise in globalization (see chapter 8). Within a decade's time, several military presidios, missions, and pueblos had been put into operation. For example, El Pueblo de San Jose de Guadalupe (present-day San Jose, the "Capitol of Silicon Valley") and the Mission Santa Clara de Asis were founded in 1777 to serve as farming outposts to supply the presidios at San Francisco and Monterey.[19] As with African slaves on the East Coast at this time, Native Ohlones were converted to Christianity. They were also forced to build the mission and the pueblo out of handmade bricks and wood.[20]

Water has played a crucial role in the history of Santa Clara County and all of California, and there have been conflicts over inadequate supply in the context of rising demand, over its centrality in major forms of production in the area, and over its contamination by different industries. The mission builders had difficult encounters with the natural environment from the very first day. El Pueblo de San Jose was built on the Guadalupe River, which flooded twice during 1777, wiping out a dam and many homes in the settlement. The flooding and droughts facing these early residents of San Jose would also make proper agricultural irrigation and the production of corn and beans nearly impossible for the first season.[21] Soon thereafter, however, the extraordinarily fertile soil provided such great crop yields that the Pueblo could stand on its own, and by 1781 San Jose's production was also fully supporting the presidios at San Francisco and Monterey.[22] This increase in production was made possible by intensive exploitation of Native labor.

Under this emerging order, on the "frontiers of Spain's New World empire the mission was the most important colonial institution,"[23] and the missions in the Bay Area depended heavily on Native American labor.[24] Thus the Ohlones played a central, albeit coerced, role in extending the Spanish empire. They labored in bondage at the Mission Santa Clara and in the Mission San Jose. Farming, bricklaying, weaving, shearing sheep, branding and slaughtering livestock, and producing lime and salt were some of the many tasks the Native folk performed. During *la matanza*—the cattle-slaughtering season—Natives would work long days producing food for the mission and tallow and hides for trade and bartering.[25] One European visitor to the Mission San Jose in 1792 later wrote, "They cultivate wheat, maize, peas and beans; the latter are produced in great variety, and the whole in greater abundance than their necessities require."[26] The production levels had to reach a scale of "greater abundance" than required because the Mission had to supply both itself and the presidios with food. This requirement for overproduction was, from the first day, an inherently anti-ecological and labor-exploitative arrangement that was bound to lead to problems. In the meantime, Native folk faced more immediate threats.

Harsh living and working conditions, disease, and Spanish assaults on the Ohlone religion and culture were routine. For example, Franciscan friars and overseers would regularly whip and imprison Ohlones for practicing their traditional religions or for not meeting production quotas.[27] Even an apologist for the Spanish Catholic Church's crimes admitted that the Franciscans "employed various forms of punishment . . . such as the stocks, the pillory, leg chains, extra work, imprisonment, or the lash. When flogging was resorted to, the maximum number of blows was limited by Spanish law to twenty-five."[28]

As with all successive systems of exploitation in the Santa Clara Valley, gender mattered a great deal.[29] Women in particular were singled out for demeaning, sadistic, and cruel treatment:

> Shocked by the partial nudity and what they took to be the "uninhibited native sexuality" of the Indians, the Franciscans inaugurated draconian measures to compel female neophytes [converts] to conform to the padres' ideals of proper female decorum and behavior. Immediately following baptism, female children over the age of five and all older unmarried females were separated from their families and locked in barracks called *monjerios*. Russian explorer Otto von Kotzebue, visiting nearby Mission Santa Clara in 1824, described one such barracks housing female Costanoan [*sic*] neophytes as a building resembling a prison, without windows.[30]

Ohlone women in the missions also experienced a significant decline in gender status. Women's status was lower in the missions than in traditional Ohlone communities for two reasons. First, unlike Ohlone social structures, the mission hierarchy of power was entirely male-dominated, supported by the Bible and other Christian teachings. Therefore, once they entered the missions, Native women were offered no opportunities to assume leadership positions comparable to the chieftainships, shamans, and other powerful ritual roles available in Native communities. "Divorce was denied them, and their role degenerated into producing children, and performing manual labor for the colonists."[31] Throughout their nearly seventy-year reign in the region, the Franciscans never appointed a single devout Ohlone convert to the status of nun. The second reason for women's decline in status was the rampant sexual assault by soldiers and padres, which violated, dehumanized, and demoralized many women in the missions.[32] As a result of higher mortality, the Native female population at the missions declined precipitously during the 1820s and 1830s.[33] Women were therefore forced into distinct roles for labor, daily behavior, and punishment in the social structure of the missions.

It is important to recognize that the Franciscan missionaries were not only doing their religious duty by converting the local Native Americans to Catholicism; they were also "pragmatic agents of Spanish colonial policy . . . who saw no inconsistency in the evangelization and exploitation of Indian converts to further colonial policy objectives."[34] And while the Ohlone were not driven to extinction, this campaign of terror amounted to genocide by any definition.[35]

Resistance

As with enslaved Native and African communities on the Atlantic coast, the Ohlones chose resistance over subjugation. These resistance efforts took a variety of forms. For example, Ohlone women in the missions were often attacked and made to perform demeaning acts for the Franciscans. As a result of this treatment and because of routine gang rapes by soldiers and frequent sexual assaults by the padres, many women committed infanticide or had abortions—both of which the Catholic Church had outlawed.

Desertion and outright rebellion were also very common in the Alta California missions. Natives were becoming quite skilled at resisting "recruitment" to the missions, which resulted in a drop in the number of neophytes the Franciscans could hold in bondage. During the spring and summer of 1795, in the wake of a major typhus epidemic, well over three hundred Ohlones and other Native persons abandoned Mission San Francisco.[36] Between 1811 and 1815, continual raids coordinated by fugitives from the missions with Natives from the Central Valley depleted the horse herd at Mission San Jose from 11,500 to just 280.[37] This had a major impact on the mission's ability to herd and slaughter cattle for the presidios and for its own residents. In the 1820s, natives like Pomponio (at Mission Dolores), the celebrated Estanislao (at Mission San Jose), and Cipriano (whose traditional name was Huhuyat, from Mission Santa Clara) led indigenous guerrilla armies and combined forces of runaway neophytes and free natives to assert their right to return to their traditional ways and defend themselves against military attacks by the Spaniards.[38] Estanislao openly defied Padre Duran of Mission San Jose with a letter that read, "We are rising in revolt. . . . We have no fear of the soldiers, for even now, they are very few, mere boys, and not even sharp shooters."[39] In 1828, Cipriano (Huhuyat) and Estanislao collaborated to lead massive runaway campaigns from missions at Santa Clara and San Jose. The Californio (Spanish-born elite) authorities launched a military campaign that ultimately crushed the rebellion but had a devastating impact on the missions' ability to survive. Throughout the 1830s, Yoscolo, an Ohlone man, raided cattle at the Mission Santa Clara and even killed Christianized Natives and others who tried to stop him. Finally, Yoscolo "led a rebellion, stealing two hundred women and several hundred head of cattle, and raiding the mission stores at will until he was caught in 1839. His head was nailed to a post near the church door as an object lesson."[40] The fact that men and women who were longtime missionary "residents" led most of these rebellions, undermined the moral position of the Franciscans, and challenged any claim that the dissent was being created by outsiders.[41]

Thus, despite being subjected to incredible assaults on their humanity, culture, religion, physical and psychological health, and environment, many Ohlones rejected and resisted this campaign at every level. However, this struggle took its toll, and the Ohlone population in northern California declined from 11,000 to just 2,000 by 1830.[42]

Still, the amount of energy and resources the Spaniards expended in the effort to control the Native resistance compromised the sustainability of the missions. The Crown soon granted land rights to colonists around the Bay Area, and those colonists took up ranching and cattle raising. This new economy grew quickly, and the trade in hides, meat, and tallow extended to the East Coast and South America.[43]

After Mexico gained its independence from Spain in 1821, ranching remained the mainstay of the California economy. Those Native Americans still enslaved at the missions were released and allowed to work as day labor on the ranches or as domestic and agricultural workers.[44] The social hierarchy that emerged was based on race and national origin, with Spanish-born elites (Californios) on top, Mexicanos in the middle, and Ohlones at the bottom.[45] This hierarchy corresponded to one's position in the local labor market and one's political power, with Spanish elites holding all formal political authority while reaping the profits from the local economies, and Native Americans working the most dangerous, unhealthy jobs and having no political rights or citizenship status.

Ecological Disruptions

In addition to the enslavement and exploitation of Native peoples, the Spaniards, Mexicans, and other European newcomers were actively taking and destroying land and various other resources from Native Americans around the Bay Area. Major ecological changes resulted. The Spaniards and Mexicans brought extensive open-range livestock grazing to California and introduced new, less productive grasses of Mediterranean origin, resulting in major damage to the native grasslands.[46] The area's ecological disruptions were linked directly to Native subjugation at the missions because when the resource base was nearly exhausted, some Ohlones, out of desperation, sought out the Franciscans for assistance.[47] The consequences were devastating for both the land and people: "[a]s agricultural laborers, missionized Indians were largely separated from the seasonal rhythms of their own food production practices, while the growth of mission farms and rangeland for cattle initiated an environmental transformation of the Bay Area and the entire coast that destroyed much of the resource base of the indigenous economy."[48]

One report found that even the padres at the missions in San Jose and Santa Clara "complained to their superiors in Mexico that the pueblo was taking too much water from the Guadalupe [river], cutting down too many trees for dams and houses and fences, and depriving the neophytes of the venerable oaks on which they depended for food [acorns]."[49]

But the missions and pueblos do not deserve all the blame. Anglo-American explorers and newcomers from Europe hunted sea otter, beaver, grizzly bear, pronghorn and tule elk, and several species of whale to near extinction by the mid-nineteenth century. The causes of this overkill shifted from an initial subsistence orientation to sport and international commerce.[50] By the 1820s, Russian explorers were exporting precious ancient redwood timber to Hawaii and elsewhere, resulting in the destruction of many old-growth forests.[51] The privatization of resources was occurring at a rapid pace. In 1847, for example, developers and private citizens exerted pressure on the City Council of San Jose to make available several thousand acres for private ownership.

Thus environmental devastation and the control, exploitation, and genocide of Native peoples were intimately bound together as the underside of the vaunted European expansion to the New World. While the destruction of flora, fauna, and landscape disturbed the Native economy, way of life, and diet, it was disease and psychological and military assaults that were responsible for the lion's share of the havoc these populations faced.[52] If environmental racism is the unequal burden of ecological hazards imposed on people of color and their surroundings, then the European conquest was the continental embodiment of this process. This was only the beginning, however, and many more groups—especially immigrants and people of color—would be swept into California's emerging multiracial society, fueled by profit, social oppression, and a reliance on a finite and fragile natural resource base.

The Gold Rush and Mining Economies, 1848–1880s

Like Silicon Valley a century and a half later, the Gold Rush was as much legend as reality. The prospect of riches and wealth beyond imagining lured thousands of immigrants to California during and after 1848, which rapidly transformed the Golden State into the most ethnically diverse state in the nation. The January 1848 "discovery" of gold by James Marshall sparked stories of California's "streams paved with gold" in newspapers all across the nation. One miner, Jasper S. Hill, wrote to his parents in January 1850, "This country is no doubt a great place to give a young man a fair start in the world as he can make money quite fast by being industrious & economical . . . in the Eldorado of the West."[53] Rumors, personal letters, and news reports told of gold lying about the countryside and underscored that the harvesting of this bounty would be the realization of the American Dream. Not only was the age-old promise of "the good life" (if one only worked hard at it) appearing in the popular press, but a more interesting notion was being put forth: that the Gold Rush was becoming a great equalizer of class relations in California. There were reports that servants were becoming masters and entire lower classes were "disappearing" as people struck it rich en masse.[54] Over the next twenty years, 300,000 whites—90 percent of them men—poured into California to pursue their dreams.

But the work of gold mining was hard, dangerous, and financially risky, and when corporations eventually took over the fields, wage labor saw steeply declining returns. The fall in wages was often dramatic, from twenty dollars per day in 1848 to sixteen dollars per day in 1850.[55] With regard to physical injuries in mining work, blisters on the hands and feet were common, as were sprained ankles and knees, crushed fingers, and "the almost universal back ache" from digging, lifting, and carrying upwards of eight hundred buckets per day (191). In addition to the myriad accidents and injuries that were a routine part of mining work, sickness was also quite widespread (186). One miner wrote of the labor involved in mining, "it would kill me, it is the hardest work that ever a man tryd to do" (190). "Mining is . . . an occupation which very much endangers health," wrote another gold seeker (138). Another miner told his family about coworkers who often exploited their bodies to the point of "prostration and sickness" (327).

If the work was difficult for these Anglo-Americans and European immigrants, times were even harder for workers of color during the Gold Rush. They faced racism that relegated them to the harshest, most undesirable and dangerous jobs. Adding to their woes, people of color were frequently mobbed, beaten, killed, overtaxed, and even banned by miners and the state government. The environmental injustices they experienced in the workplace of the gold fields were occurring simultaneously with the catastrophic impact of the Gold Rush on California's ecosystem.

The Importance of Santa Clara County to California's Gold Rush

While the majority of mining took place further north, Santa Clara Valley (where Silicon Valley is today) played a central role in the California Gold Rush. At the time, San Francisco and San Jose were the two largest cities in the state. As the Gold Rush grew in its importance to the state economy, national and multinational corporations became involved. The first such mining corporation to form during the 1850s was the California State Mining & Smelting Company, based in Santa Clara County.[56] More importantly, the discovery of the Western Hemisphere's largest quicksilver (mercury) mine, just twelve miles south of San Jose, actually made the Gold Rush possible in California and facilitated gold and silver mining throughout the Americas.[57] In 1845, Captain Andres Castillero, a young Mexican cavalry officer, was awed to discover that the cinnabar clay at a famous vermillion cave in the area contained quicksilver. Quickly filing a claim on the property, he named it the New Almaden Mine.[58]

Quicksilver was an essential element in the process of extracting gold and silver from raw ore during this era. By using quicksilver, miners were able to amalgamate finer gold particles, "flour gold," or dust that otherwise got washed away. This was an ancient technique introduced in Mexico by the Spaniards during the sixteenth century, and brought to California by Mexican and Russian miners in the nineteenth century.[59] "Spanish Town" (the segregated Mexican quarters) quickly developed around the New Almaden mine, and by 1850 it was running at full production, employing nearly two thousand mostly Mexican workers. Although many Mexican miners were skilled (having previously worked claims in the state of Sonora), Cornish immigrant miners had more experience at deep shaft mining and were brought into New Almaden to do this type of work after the surface deposits had become exhausted. Most of the time, however, Mexicans, Chinese, and Native Americans (children included) would be used for the most dangerous mining work. These are only some of the many examples of the racial division of labor that occurred, wherein different ethnic groups were used to work jobs with varying degrees of risk, thus sustaining a wage and task hierarchy that depressed all groups' wages and generally preempted any collective solidarity.[60]

The racial division of labor and of everyday activities had real consequences for different racial and ethnic groups. In the New Almaden mine, Anglo-American and European workers were employed in the less taxing occupation of "reduction" (extracting quicksilver from ore) and were paid $5–$7 per day. In contrast, Chinese and Mexican workers were employed in the much heavier and less desirable occupation of "ore-carrier," and were paid $2–$3 per day.[61]

Starting in the pit of the mine, the ore-carriers would fill a large sack or pannier made of hide with two hundred pounds of ore and then ascend the *escalera* (the ladder-like circular path) to the surface. Open at the top, the pannier was flung over the shoulder and supported by a strap that passed over the shoulders and around the forehead. Ore-carriers made from twenty to thirty trips a day up the *escalera,* for all the ore was carried to the surface by hand. The *escalera* was narrow, slippery, and lighted only by a few flickering torches. Visitors told of seeing the *tanateros* (ore-carriers), dressed in pantaloons rolled tight above the knees, and calico shirts, hurrying up the *escalera* with "straining nerves and quivering muscles."'[62]

At various times, Native Americans, particularly children, were also employed in the quicksilver mine, as they were perhaps the most expendable and least valued human beings in the state of California.[63] In addition to the harsh racial division of labor and wages, the living quarters, churches, schools, hospitals, and benevolent societies around the New Almaden mine were also segregated by race.[64]

Moreover, quicksilver mining exposed workers to mercury, a toxic substance that can cause a range of fatal illnesses.[65] This was an early form of chemical exposure in an area of California that would not become infamous for the prevalence of toxic work until one hundred thirty years later, when Silicon Valley's electronics firms introduced new forms of hazards during their very own "Gold Rush."

San Jose and Santa Clara County Face Early Environmental Challenges

The 1848 annexation of California by the United States brought about a change in the ownership of the mines (from Mexican to U.S. and British owners) and all common lands (*ejidos*) in El Pueblo de San Jose were sold off to private landholders. At the time, the Valley's economy was largely based on ranching, with the main exports being wheat, tallow, hides, and quicksilver.[66] Santa Clara Valley was becoming a major economic center in the region, but it had a few setbacks, such as the floods of 1849, which made a virtual lake out of the transportation route between San Jose and its major port at Alviso.[67] This was yet another example of how water played a role in the Valley's trials and triumphs.

In 1850, El Pueblo de San Jose became the City of San Jose. California achieved statehood that September, and San Jose was soon named the capitol. Within the first two years, the city council was compelled to address environmental concerns:

> Although water was plentiful, surface contamination was a serious problem. Water was supplied by the Mexican-period water ditch (*acequia*) that ran west of Market Street, parallel to the Guadalupe River, and by shallow wells, six to 10 feet deep. The water ditch had accumulated trash and garbage, and several ordinances were passed to regulate its use. Following the Mexican procedure, a ditch overseer was appointed, and adjacent property owners were made responsible for the

ditch and maintenance of its banks. It became a misdemeanor to throw filth into the ditch, to wash clothing in it, or to impede the passage of water.[68]

Acequia technology and the associated social-legal traditions to protect it were borrowed from Mexico. Predictably, however, the adoption of the *acequia* practices was soon trumped by the founding of the San Jose Water Company, which was established in 1852 and shortly obtained the exclusive right to sell water to city residents.[69] Thus, in a pattern reminiscent of when the Spaniards took Ohlone land and natural resources and privatized them, Anglo-Americans carried on this tradition a century later with water, land, minerals, and livestock. Water more than anything else was both the lifeblood and the Achilles heel of the Santa Clara Valley and California economies.[70]

Social Hierarchies on the Frontier: Mexicans and African Americans

In 1850, California's gold fields were still considered "undeveloped" or "frontier" territory, attracting fortune seekers without families. That was one reason women accounted for less than one-tenth of the state's population and were only 3 percent of the population in mining counties. The absence of women forced men to learn a wide range of domestic skills in order to survive.[71] The women who were in the gold rush counties, however, also broadened their occupational skills within and beyond the domestic sphere, including the job of schoolteacher, baker, shopkeeper, boarding-house operator, and of course, gold miner.[72]

With the Treaty of Guadalupe Hidalgo ending the U.S.-Mexican War in 1848, the United States annexed all of California and most of the southwest. This was a "land grab" on a colossal scale. In one of many efforts, business and white working class interests succeeded in passing the Land Law of 1851, which made all Spanish and Mexican land grants subject to review and rejection.[73] This law quickly opened the door to a flood of legal challenges, the dismantling of the Californio ranching economy, and the transfer of massive landholdings to whites. Just eight years after the United States signed the treaty with Mexico, the majority of land guaranteed to Californios had been taken.[74] The fertile land of northern California had switched hands first from the indigenous

Ohlones to Spain, then from Spain to Mexico, and then from Mexico to the United States.

This era saw the creation of a new racial hierarchy, with Mexicans (who were eventually allowed citizenship rights) just below whites; next came African Americans, then Asians, and then Native Americans.[75] Whites allowed Mexicans a relatively elevated status because most of them were Catholic (a European-based religion), because they were Spanish-speaking (a European language), and because, as *mestizos*, they had physical features and bloodlines that were part European.[76] Thus, they were closest to Anglo and other European immigrants in appearance, language, and culture.

The U.S. annexation of California displaced many local Mexicans out of ranching and pushed them into mining. Other Mexicans (and Native Americans) went to the mines when their employers—owners of large land grants in the interior valleys—brought them to work the claims.[77] Still others, who moved northward to the gold mines during 1849 and afterward from "Old Mexico," were like many Chinese miners in that they were experienced excavators (from the state of Sonora). This expertise was both a curse and a blessing in that, while they were more efficient at extracting gold, their mining acumen gave them an advantage white miners would soon come to resent. Mexican miners were essential to the success of many gold mining companies. The Morgan Hill mine in Calaveras County was originally worked by Mexicans and eventually yielded more than two million dollars in gold.[78] Mexicans taught Anglos the techniques required for placer mining and quartz mining.[79] After Mexicans had imparted their skills they were viewed as impediments to the success of white miners, so as early as 1850 there were systematic attempts to remove Mexican miners by "rounding them up," imposing fines to be paid in gold, and launching violent attacks against them.[80]

African Americans, although mostly enslaved at the time, were next in the social order because they spoke English, practiced Christianity, and had assimilated many European cultural patterns. Despite efforts to exclude them, African Americans, both free and enslaved, were present in California during the Gold Rush as well. Many worked the mines as slaves, while others who were instrumental in working high-yield claims for their masters were able to purchase their freedom. The white resentment against African Americans was considerable, and they were easy targets for lynching, robbery, and mob violence. The estimated 1,500 African Americans who worked the gold fields often did so at their peril.

They labored in places that were given names like Negro Hill, Negro Bar, and Negro Flat—few of which were ever documented on official maps.[81]

Very few African Americans settled in Santa Clara County or San Jose at the time. Those who were in the area held occupations as slaves, laborers, barbers, farmers, and teachers. The African American community built its own church and public school, as segregation was a way of life. In many ways, African Americans served as a model for how the Anglos treated the Chinese. The denial of citizenship rights, the exclusion from court proceedings, and the lynchings directed at Chinese workers and residents were practices that had first been used against African Americans.

Chinese Immigrants and Native Americans

Popular culture regarded Asians as exotic pagans with totally incomprehensible cultural, linguistic, and religious practices. Whether it was food, dress, or social rituals, the "heathen chinee" and the Japanese "yellow peril" were unambiguously deemed alien and nonwhite. For these reasons they were second to last in the social pecking order at the time. Thousands of Chinese men immigrated to the United States during the late 1840s, coming mainly from Guangdong Province, where a major flood led to a collapse of the regional economy.[82] They sought a better life and wealth in the gold fields of northern California. For this reason, many Chinese referred to the United States as "Gold Mountain." The U.S. government and industry viewed this wave of immigration as an asset to the mining sector because many Chinese men were skilled in excavation techniques. By 1860 Chinese immigrants made up nearly 10 percent of California's population and 25 percent of its labor force, and accounted for 35,000 gainfully employed miners in the state.[83] Because of their exceptional excavation skills, Chinese workers were often employed in "the dregs," where they searched for gold in mines that had already been exhausted by independent white miners. This work was both especially dangerous *and* socially undesirable, as it required doing the risky and dirty work that most white miners would not or could not do.

During the Gold Rush, the California white working class (and some business interests) almost immediately opposed Chinese immigration. The white Anglo and European immigrant majority were organizing a movement whose goal was to maintain California as a "free labor" state.[84] The driving principle behind this movement was that the white working class could only achieve upward mobility and economic independence in a capitalist society free of slavery. Slavery and other forms of bonded labor were generally associated with nonwhite persons, so a "free labor" movement was also an effort to restrict the influx of people of color into the state.[85] Because the majority of Chinese immigrants to the Americas were conscripted into indentured servitude or some other form of bonded labor, this movement was directed primarily at them.[86] Vigilante murders and mob violence by whites were leading factors in driving away many Chinese workers, but other means were equally effective, including the Foreign Miner's Tax—a state law passed in 1852 requiring all non-citizen miners to secure a license to mine and to pay a tax of three dollars per month. Chinese immigrants were deemed "nonwhite" by law, thus rendering them ineligible for citizenship.[87] The anti-Chinese movement of the nineteenth century was punctuated most forcefully by the passage of the Chinese Exclusion Act of 1882, which terminated all Chinese immigration. By this time more than 1,200 Chinese men had died as a result of the illness, overwork, explosions, and avalanches they faced while building the transcontinental railroad, for which work they were paid only two-thirds what their white coworkers made.[88] The Chinese resisted every one of these discriminatory laws.

Completion of the transcontinental railroad in 1869 enabled Santa Clara County to become the leading fruit producer in the nation, because the railroad allowed growers to access markets for its produce, much of which would otherwise rot on the ground during years of overproduction.[89] The railroad also brought many Chinese to California, including the Santa Clara Valley, from points east. While there were fewer than ten Chinese residents in the Santa Clara Valley during the first days of the Gold Rush, the number increased significantly when angry white miners drove the Chinese out of the mines. They worked in and operated laundries, restaurants, and prostitution houses in San Jose during this time. Chinese workers were vital to construction of the railroad and to the subsequent growth of Santa Clara's agriculture because they provided cheap labor in the fields.

At the base of the new social order in the post-1848 era stood the Native American population. The status of Native Americans in the United States was clear: they had no citizenship rights and remained at the bottom of yet another social hierarchy. Whites generally believed them to be an unclean, physically unattractive race with no redeeming qualities in their history or culture. As residents and caretakers of the land that white capitalists and Anglo/European settlers coveted so dearly, they were viewed as obstacles to progress. In other words, there was no place for them in the future of California society. They were *personae non gratae*. California's Native peoples were the primary victims of the Gold Rush, suffering forced removal, theft of gold-rich lands, epidemics of new diseases brought with the Anglos, and scores of sanctioned and unofficial massacres of "hostiles." This was genocide in progress. Between 1848 and 1868 the Native American population in the State of California declined from 170,000 to 50,000; in the Santa Clara County area the population dwindled from 4,000 to an estimated 100 persons, chiefly as a result of disease.[90] An especially noteworthy example of how the genocide of the Ohlones and the ecocide of Santa Clara Valley's natural resources were linked was the case of Chief Lope Inigo (1760–1864). In 1844, Inigo was granted 1,600 acres of land, one of seven grants given to Natives by Mexico after the missions were disbanded. The land was called Rancho Posolmi, a Native American name whose meaning is lost. It was also called Pozitas de las Animas, or Little Wells of the Souls, after a spring that flowed for some years on the property. Chief Inigo was one of the few remaining members of his tribe, and after his death in 1864, whites, who had illegally squatted on his land, took it over. The Holthouse family farmed the land and sold fresh peas under the brand name "Ynigo," with a picture of an Indian Chief on the package. In 1933 the land was "developed" into a Naval Air Base—Moffett Field, in Mountain View. In the 1980s, this base was declared a Superfund site so toxic that the United States Environmental Protection Agency has estimated it would take three hundred years to clean up.

Inigo, an Ohlone man and a member of a nation that suffered from a most brutal genocide, was thus "given" land—which had first been stolen from his people—by the Mexican government. Then, that same land was taken back after his death and used to market produce under his name (without his permission), and finally to house weapons of mass destruction that harmed the soil and drinking water in the area to the point that a superficial cleanup would take centuries. This case represents an egregious but illustrative example of the desecration of both Native Americans and their land throughout each stage of the Santa Clara Valley's history.[91]

Impacts on the Natural Environment

As we showed earlier, the South San Francisco Bay Area was a place of great natural beauty. But, like the Spaniards before them, the thousands of miners and capitalists who entered California's gold fields in droves approached the ecosystem with little care:

> There was no obvious concern for the environment. Anything that stood in the way of the gold seeker was pushed aside or destroyed, whether a grizzly bear or a mountain. Ruthless exploitation with no thought for tomorrow was the basis for the way of life in gold-rush times.[92]

This lack of environmental awareness was supported by a strong sense of Manifest Destiny and a Lockean ideology about the American West, which entailed "ignoring all prior Indian land use and property rights."[93] These sentiments were complemented by a religious undergirding that viewed U.S. expansion as a "God-given" right that had been "earned" through the just defeat of Native American nations and of Mexico.[94] Punctuating the damage wrought since the dawn of the Spanish arrival nearly a century earlier, the Gold Rush reinforced the ideology that natural resources existed only for the colonizers' personal profit and that indigenous peoples and people of color—particularly immigrants—were expendable means to these economic ends.

The Santa Clara Valley area was pivotal to the success of the Gold Rush mainly because of the New Almaden mine's location south of El Pueblo de San Jose. Quicksilver was the New Almaden's export, a substance that is extremely harmful to human health and that compromised the well-being of the workers who extracted, processed, and applied the "quick" to the ore. Quicksilver/mercury is also renowned for destroying fisheries and other key components of entire ecosystems. Not only is it a powerful toxin, but it bioaccumulates throughout the food chain, causing harm to multiple organisms.[95] Significant volumes of quicksilver were used and dumped (usually lost) into California rivers during the Gold Rush. In one case, at the North Bloomfield mine, the company working there dumped or lost 3,798 pounds of quicksilver into a river over a two-year period.[96] The historical origins and contemporary legacy of environmental racism in both the workplace and the communities of the Bay Area are illustrated in the fact that 75 percent of the people who fish the San Francisco Bay for food today are people of color (most of them Laotian, African American, and Vietnamese) and they ingest with their seafood mercury that is a century and a half old, courtesy of the Gold Rush.[97] Furthermore, the Guadalupe River watershed, the very same river on which El Pueblo de San Jose de Guadalupe was founded in 1777, is today the most mercury-contaminated river basin in the United States.[98] In addition to quicksilver, various acids were routinely used to separate gold from silver in the mines, and cyanide was used as a quicksilver substitute beginning in 1890.[99]

In the decades after the Gold Rush and prior to the turn of the century, mining companies continued to push throughout California's interior for valuable minerals. Most of this work was becoming heavily capitalized and involved much more expensive technology than before. Unfortunately, this generally meant more "efficient" mining achieved through greater destruction of the landscape. The most egregious example was the growth of hydraulic mining, which involved pumping millions of gallons of water through hoses aimed at mountain sides, removing trees and masses of earth to expose potential veins of gold.[100] This method of extraction "reveals more perfectly than almost any other the hot, sweaty, rip-it-up, what-the-hell enthusiasm and carelessness of those years when men created waterfalls and hooked up pipes and monitors to wash away hills, in the great business of hydraulic mining."[101] One observer wrote, "[I]t is impossible to conceive of anything more desolate, more utterly forbidding, than a region which has been subjected to this hydraulic mining treatment."[102]

Hydraulic and other forms of mining revealed an inescapable environmental fact: the most valuable natural resource commodities in California—other than gold—were toxic quicksilver, cyanide, and precious, finite water. Used during every step of the gold mining process, water's value as measured by demand rose exponentially as more miners entered the field and the markets for gold expanded. "Mining and water became synonymous."[103] The need for water in the mining process was a ready-made opportunity for corporations to control the gold fields *and* the workers who labored in them. Corporations quickly began acquiring "prior appropriation" rights to water, allowing them the exclusive privilege to exploit certain bodies of water.

Early on, it was clear that the returns from mining were diminishing and a new economy would have to take the place of the old. Business leaders and speculators were turning to California's "new gold"—wheat. Farmers soon discovered that the sale of wheat produced in the state's fertile soil was yielding higher profits than gold.[104] Farming was already the occupation of choice for most Californians, especially for the residents of the Santa Clara Valley. During the 1850s, Santa Clara County's farmers produced more than 40 percent of the state's wheat, the largest wheat crop in California.[105] However, the mining companies still in operation continued using hydraulics in the highlands and mountains, and this created epic conflicts with farming interests. During the 1870s and 1880s, farmers were angered at the spoiling of their lands by millions of gallons of debris, tailings, and silt from mining operations miles away. Political and military struggles ensued, with the development of Anti-Debris Associations, which were among the first environmental organizations in the United States. These groups were militia-like organizations set up to combat the mining industry and its pollution of farmland.[106] In 1882, a judge handed down a decision against the North Bloomfield Gravel and Mining Company restraining them from dumping debris or tailings in rivers. The industry challenged the ruling, but in 1884, the U.S. Circuit Court in San Francisco issued a perpetual injunction against hydraulic mining—a death blow to the industry and a victory for the farmers and Anti-Debris Associations that ushered in the new agricultural economy in California.[107]

Conclusion

The harms done to California's ecosystems and its less powerful peoples were not inflicted primarily by heartless individual miners eking out a meager existence. The major driver behind the Gold Rush and the damage done to people and the environment was the corporation.[108] Corporate power won out over individual and collective rights for workers and different ethnic groups. Corporations cornered the market on quicksilver and gold, manipulated share prices, cheated investors, and drained smaller companies. They were also chiefly responsible for the untold injuries inflicted upon California's natural environment, its Native peoples, and the miners themselves, many of whom were immigrants and people of color. Ultimately, large corporations were responsible for doing great damage to California's economy as well.[109]

The Gold Rush's significance is captured in the exponential increase in ecological disruption committed in the name of building the state's economy, the continued volatility in racial/ethnic relations in communities and workplaces, the rise of the multinational corporation, and the hyper-exploitation of the working class, particularly immigrants and people of color, via their concentration in high-risk, low-wage jobs. Santa Clara County played a central role in this process in that this was the location of the Western Hemisphere's largest quicksilver mine, without which the Gold Rush would not have happened, and this was where the first large gold mining corporation began its operations.

The irony of environmental destruction in the Santa Clara Valley is that sustainable options were always available to the Spanish, Mexican, or Anglo establishments (depending on the historical period in question). For example, the Ohlone's sustainable fishing, farming, and mining practices were proven, but were completely disregarded by their conquerors in favor of overharvesting, overproduction, and depletion and despoiling of marine life, agriculture, and silver and gold. The Spanish-Mexican practice of building communal *acequias* was also pushed aside in San Jose and elsewhere in favor of the development of corporations that privatized, owned, and sold water. The use of the *fong sei* by Chinese miners is another example. The *fong sei* was a place outside the mining camps where Chinese workers would gather to be at peace in unfettered natural surroundings, such as trees, streams, and unmined earth. This was their way of reaffirming nature's inherent worth and their refusal to spoil at least some of the land, no matter how much gold or other resources might lie within.[110] Like the people who practiced these traditions, the Ohlone way of life, the *acequias*, and the *fong sei* were all ignored and trumped by the corporate form of development that viewed natural resources and people of color and immigrants only as a means to a private, profitable end.

In the 1880s, the state of California turned once again toward agriculture, and the Santa Clara Valley's newest economy would, for the second time in a generation, place it at the center of national and international markets for its products.

3

The Valley of the Heart's Delight

Santa Clara County's Agricultural Period, 1870–1970

Introduction: Orchards and Canneries in the "Fruit Bowl of America"

Wheat production in the Santa Clara Valley peaked in the early 1880s when mechanization and year-round harvesting led to poor soil conservation and overproduction. Wheat was rapidly becoming a crop of the past, and Santa Clara County's fruit and vegetable processing industry had been competing for dominance since its start in the 1870s. Like the Spanish missionaries a century before, Italian, French, and German immigrants brought cuttings of vines and fruit trees to Santa Clara County from their home countries, and these seeds laid the foundation for the next stage of economic "development" in the area.[1] The demand for fresh and packed fruit rose exponentially during and after the Gold Rush and the Civil War. The Gold Rush "ushered in wildly inflated prices for fruit" and was responsible for exponential population growth in the state, creating an opportunity for Santa Clara County to profit from its fertile soil.[2] Santa Clara County became the state's leading producer of fresh, dried, and canned fruit, and the nation's leading producer of prunes, which could easily be shipped by rail to markets on the East Coast. As late as the 1950s and early 1960s, some observers claimed that the county was the single largest producer of canned produce in the world.[3]

Immigrants, People of Color, and Women Workers in the Valley

The need for cheap and compliant labor in the orchards and canneries of the Valley and the Bay Area motivated producers to encourage an influx of immigrant workers to the region. The majority of these new workers arriving in California around the turn of the century were Chinese, and later Japanese.[4] After being driven out of the mining camps, many male Chinese workers were employed in the canneries of San Francisco, a burgeoning industry looking for cheap labor. The majority of Chinese residents in San Jose were displaced gold mine or railroad workers.[5] Most of these 2,700 Chinese residents began sharecropping and performing stoop labor in the strawberry fields of Santa Clara County. They received a half share of their earnings from each harvest. The land where these Chinese laborers worked was the undesirable marshy lowland area near the Bay (around Alviso), where they labored to reclaim wasteland and plant and harvest the berry crop.[6] Between 1870 and 1895, San Jose's Chinatown was burned down by angry whites two times and rebuilt by Chinese residents each time. Thus, these Chinese immigrant residents of San Jose lived and worked under harsh conditions. This was a classic example of environmental racism: this group of immigrants (who were also people of color) was forced to labor and live in environmentally unhealthy areas, with little control over the natural resources around them, on which they toiled for the profit of white entrepreneurs.

By 1900, many Japanese immigrants had settled in San Jose and were living in a farming cluster in the Alviso area, previously worked and occupied by many Chinese.[7] Many of these men were married, and their families were able to achieve greater mobility than the Chinese single male workers of the previous generation. The reason for this was the presence of Japanese women, whose labor was fundamental to the family's survival and well-being.[8] However, Japanese workers soon became militant in pursuit of their rights and conducted many strikes after 1900.[9] Like the Chinese, they were immediately targeted for immigration restrictions under the Gentlemen's Agreement of 1906.[10] A number of Japanese families also succeeded in purchasing their own land, and the reaction by whites was predictable. Compounding the Gentlemen's Agreement, Congress passed the Alien Land Act of 1913 (reenacted in 1919) and the federal restriction on further Japanese immigration in 1924. The Alien Land Act made it illegal for non-citizens to purchase land for farming, and like the Foreign Miner's Tax Law of the 1850s, this legislation also targeted Asian immigrants because they were still denied eligibility for citizenship.

More than seventy years after the first wave of Chinese workers arrived in California, anti-Asian sentiment still held sway. Testifying before a Congressional committee in 1925, S. Parker Friselle, a California rancher, complained that the Japanese "do not want to work for anybody else. The Japanese wants to work his own farm." Another critic wrote a year later, "If he [the Japanese] could have been kept as a laborer, it would have been of great benefit to California."[11] The instant the Japanese sought upward mobility and access to the rights European immigrants were formally granted, they were rebuffed.

Women workers were highly sought after because they could be paid less and employers believed that they had greater manual dexterity—a plus for handling delicate produce.[12] The following are newspaper advertisements that were used to recruit women and girls to work in Santa Clara Valley canneries during the 1910s:

> WOMEN AND GIRLS—Do you want to spend two months in the country and earn good wages, with steady work in a fruit cannery? Tents furnished. Working now.
>
> WANTED—Women and girls for cannery work in country: commencing immediately and steady until November 1; no experience necessary; good wages. Cottages and tents furnished. Call at once.[13]

Not only women, but girls aged twelve and above were recruited to do piecework in the canneries. Employers also sought out immigrant female workers because they were believed to be much easier to control than native-born women. Nearly 90 percent of these women workers were of foreign birth.

No census information is available on Santa Clara County's Latino population until 1900, and even for that year it is quite incomplete. After World War I and around the time of the 1924 Johnson Reed Act (restricting immigration from southern and eastern Europe, with explicitly racist overtones directed at these darker-skinned populations), the United States Food Administration aid program for Europe brought about a significant national increase in the demand for Santa Clara Valley fruit. This program was a boon to growers in the area, who immediately turned to Mexico for a new reserve army of cheap labor. During the 1930s, thousands of Mexicans emigrated to California to work in agriculture. Santa Clara County hosted many of these migrants. Throughout the 1930s in the Valley, the lion's share of hard, migratory farm labor was done by Mexicans, Filipinos, southern Europeans, and working poor Anglo-Americans, many of whom were refugees from the Dust Bowl in the Southwest.[14] These "pickers" spent several months out of the year living in makeshift tents and fruit-tray shelters, enduring very difficult working conditions.

The Mexican migrant population also increased as European immigrants and Anglo-Americans left farm labor to move into the canneries. The growth of the Mexican immigrant population continued throughout the twentieth century, increasing from around 15,000 people in Santa Clara County in 1970 (the largest group of foreign-born residents in the area)[15] to nearly 227,000 in 1980.[16] By 1995, the total Latino population constituted 23 percent of the county's residents.[17]

Changes in immigration law dramatically altered the ethnic composition of the United States in general and of California and Santa Clara County in particular. The Immigration Acts of 1924 and 1952 created and maintained a system of national origins quotas—a racist practice that tended to favor western and northern Europeans over all other groups. The Immigration Act of 1965 was passed with the intention of undoing the historic racial biases against eastern and southern Europeans by allowing more of them in. However, this law had the unintended effect of allowing more Latin American and Asian immigrants in because the emigration drivers from within Europe had slowed down by that time.[18] Since 1965, the majority of immigrants to the United States have been people of color, and Santa Clara County's demography reflects these changes as well, with whites now making up less than 53 percent of the population.

Labor Unrest: Unions Make a Stand

Agricultural work in California—whether on farms or in canneries—was grueling labor that fell disproportionately to immigrants, people of color, and women. One scholar commented on California's agricultural labor system, "For over a half century this sordid business of race exploitation has been going on in the state and it would be difficult to find a meaner record of exploitation in the history of American industry."[19]

The cannery and farm field workers in the Santa Clara Valley were organized in one form or another beginning in the 1910s. Ironically, from the start, much of the labor strife was directed internally: at various unions and their leadership, rather than at the management or ownership of the orchards and canneries.[20]

While farm workers were mostly people of color at the turn of the century, Portuguese and Italian immigrant workers (mostly women) dominated the cannery workforces. During the World War I period, the canneries in Santa Clara County were making efforts to increase productivity by lengthening the work day and reducing wages—changes that disproportionately impacted the most recent (and poorest) immigrants. These modifications of the work environment heightened tensions between labor and capital, which were also ethnic tensions, as the owners were largely Anglo-Americans. Over the next fifty years major strikes and labor conflicts broke out between agricultural unions and management, with the state often intervening with police and National Guard actions. Piece-rate practices, low wages, poor working conditions, racially biased pay scales, and battles over union control were the major points of contention. Unions involved over this time period included the International Workers of the World (IWW or the "wobblies"), the Cannery and Agricultural Workers Industrial Union (CAWIU), the Teamsters, the American Federation of Labor's California Council of Cannery Workers, the United Farm Workers, and the CIO-affiliated United Cannery, Agricultural, Packing, and Allied Workers of America (UCAPAWA). The various racial and ethnic groups (mostly women) working in the fields and canneries from the 1920s to the 1970s included Chinese, Japanese, Mexican/Chicano, Puerto Rican, Filipino, Portuguese, Italian, Spanish, and Anglo.

The Environmental and Social Impacts of Canning: Environmental Justice Concerns on the Job and in the Community

Aside from the persistence of racial and gender domination of labor, basic environmental justice concerns on the farms and in the canneries were rampant and intense. The Mexican, Filipino, and European immigrant "pickers" working the farms during the 1930s confronted many challenges: "The pickers suffered sore knees, sunburn, and backaches from stooping to pick the fruit off the ground. Migrant families often had several children with them who picked fruit as well. There were no restroom facilities, and the only drinking water was what the family brought to the orchard in glass jars."[21]

Additionally, the pesticide use and exposure in the industry after World War II contributed to an average of 1,000 farm workers killed and more than 300,000 people taken ill each year.[22] The arrival of mechanization in the industry in the 1940s saw a sharp rise in the rate of occupational injuries. Many women workers were laboring at increased speeds to take home a decent wage under piece-rate conditions, and this often resulted in severed fingers, strained tendons, and increased psychological stress. Injury rates hit their highest point in 1942 during the frenzy of wartime production.[23] Piecework in particular, and poverty wages in general, forced many workers to recruit their children to the job. Like the piece-rate work that prevailed in Silicon Valley's electronics industries many years later (see chapter 7), several forces were driving this phenomenon. First, of course, the cannery owners and managers sought out child labor because it was even cheaper than adult female labor. Second, parents often wanted their children working in the canneries because child labor could mean the difference between "making ends meet" and going hungry:

> The canning industry of California finds a use for employees of all ages from twelve up. . . . The child labor law of California does not permit children under twelve to work. It does, however, give children between the ages of twelve and sixteen the right to work provided they obtain permits to do so from the board of trustees of their school district. The canneries of the Santa Clara Valley do not employ as many boys as they used to as the various managements feel that they are too playful. There are, however, a considerable number of girls over the age of twelve employed in the preparation of the fruit. They are more docile and easily managed than the boys and they are more physically fit to perform the same light, rapid work that the adult women are expected to perform. . . . There is of course pressure brought to bear on the cannery managements by the employees with large families to employ their children. This is especially noticeable among the foreign workers who seem to regard their large families as an economic asset.[24]

Piecework can create a sense that employees are working for themselves, with the result that many engage in "self-exploitation" and work as hard as possible. This is what one sociologist has termed "manufacturing consent"—that is, the consent workers give to management to allow themselves to be exploited above and beyond what would seem necessary to keep one's job and to maintain production flows.[25]

The canneries of Santa Clara County were notoriously dirty and unhealthy. One foreman of the mincemeat room in a prominent cannery during the 1920s told a researcher that "a certain amount of dirt, taken internally, was good for one."[26] However, even that researcher was a product of his times and failed to question the racial division of labor that concentrated immigrant women in the most dangerous job, that of "cutter," where injuries were quite frequent:

> There are a great variety of types of women to be found in a cannery. They vary along racial, temperamental, and cultural lines. To a certain extent these women are segregated by the ordinary processes of division of labor, which is so characteristic of all present day industry. For instance, the cutters, whose work requires less intelligence than that of the canners, may be composed of foreigners or of those of less education and culture than that possessed by the canners.[27]

In the late 1970s, when sociologist Patricia Zavella was researching her book, *Women's Work and Chicano Families*, cannery workers told of their experiences with exhaustion, dehydration, swollen feet, dizziness, headaches, and varicose veins as a result of the labor process. Extreme heat and cold, excess humidity, inadequate ventilation, and noise pollution were also common problems.[28] From 1958 to 1970, canneries in the United States reported a work-injury rate nine points higher on average than any other manufacturing industry.[29] In the state of California in 1976, food and related industries reported the second highest rate of work-related injuries. The canned fruit and vegetable sectors in California reported on average eighteen injuries and illnesses per one hundred workers, while *all* California industries reported an average of just ten.[30]

As with the mining industry that preceded it and the electronics industry that followed it, the cannery was also a site of toxic chemical exposure and chemical burns. The mostly Chicana/o workforce routinely had to handle caustic, chlorine-based, and lye solutions while processing produce. Workers reported that the fumes from the lye often made them nauseous and resulted in absences from work.[31] In violation of environmental and occupational health laws, there were no Spanish-language warnings regarding the nature of chemical dangers or of chemical use in these workplaces.[32] At one San Jose plant in 1978 seven women workers were hospitalized after a chlorine spill; that same year, workers filed a complaint with the California Occupational Safety and Health Administration charging canners with allowing excessive noise levels.[33] Finally, as was the case with a range of other work-related issues, the union was blamed for its nonchalance and neglect with regard to health and safety. From both management *and* the union, women cannery workers of color faced psychological, physical, economic, environmental, and cultural threats to their health, safety, and well-being.

The home environment, neighborhoods, and barrios during the cannery era also faced a range of negative impacts. The barrio of Sal si Puedes, where Mexican immigrant and Chicano workers were concentrated, experienced infamous sewage problems and was known for its low quality housing stock. Disease, injury, and sickness were also rampant during the 1950s, as a result of poor working conditions, environmental degradation in the community, and poverty.[34] Julio, a Mexicano resident of this barrio, told one researcher, "we would like to live in a house with a good floor and a tight roof than in a shack; we like to eat meat when we can buy it. But sometimes we can't."[35] With regard to city services, Sal si Puedes was notoriously underserved, with no curbs, gutters, storm sewers or sidewalks.[36] The *environment* is where we "live, work, and play," and San Jose's Mexican immigrant and Chicano community faced environmental degradation in all three of these arenas.

The Agricultural Industries' Broader Ecological Impacts on Santa Clara Valley

The Santa Clara Valley was unparalleled in its visual appeal, with orchards blanketing a significant portion of the area between 1880 and 1950.[37] The Scottish immigrant and founder of the Sierra Club John Muir was taken with the Valley's scenery. He wrote, "Its rich bottoms are filled with wheat fields, and orchards, and vineyards, and alfalfa meadows."[38] Elizabeth Nichols, a Communist Party organizer who worked in the canneries of San Jose during the Depression, later reminisced about the glorious landscape in those years: "You really have to have seen it. You can't possibly imagine how beautiful it was to look down from a hillside to a carpet of white blossoms."[39] A geographer described Santa Clara Valley's beauty during the 1930s: "From the west the valley floor seems completely carpeted with orchard."[40] None of this new economy would have been possible had it not been for the presence of some of the world's most fertile and productive soil and major water aquifers below that surface—both of which were being taxed beyond their limits.

As with gold processing more than eight decades earlier, the orchard economy in Santa Clara relied heavily—too heavily in fact—on water, which "became a treasured commodity."[41] Water was the one commodity northern California industry seemed to consume and pollute without end, and it played a significant role in the emergence of Santa Clara County as an agricultural giant.[42]

San Jose was the only major California city without a Sierra water source. Almost all of the water used for irrigating orchards and croplands was drawn from wells and aquifers. And like any economy experiencing rapid growth, Santa Clara County's demand for this natural resource was growing. From 1915 to 1933, the pumping draft increased from 25,000 to 134,000 acre-feet per year, resulting in a significant drain on the ground water supply—a drop of 95 feet. Between 1915 and 1967 the land on which the Alviso community stood sank by 12.7 feet, a direct result of ground water depletion. Through the mid-1930s, almost 90 percent of underground water was used by the agricultural sector.[43] As we saw with mining technology during the Gold Rush, water pumping machinery became more sophisticated and expensive, and was in ever greater demand, as the water table receded even further.[44] Water pumping technology also became more environmentally destructive; not only did it deplete ever more thoroughly the aquifers and wells, but the very power sources that allowed the pumps to function became more resource intensive, changing from wind, to steam, to gasoline and electric motors of ten, twenty-five, and seventy-five horsepower.[45]

The City of San Jose and the County of Santa Clara worked diligently to study the water problem and devise methods of preventing floods *and* conserving existing water reserves. Voters cast ballots in favor of constructing a new reservoir in the early 1930s. The Santa Clara Water District used this mandate to build several reservoirs, or "percolation ponds," which allowed water to spread and be absorbed into underground aquifers for storage. This plan may have saved the county's economy, because while the water table was drained to an all-time low in 1934 (exceeded in 1948), the next year the table rose ten feet as a result of the construction of five reservoirs. Water scarcity produced a lot of headaches in Santa Clara County, but as in times before, it also presented an opportunity for capitalists to achieve big gains. While Santa Clara County saw growth in the orchards and canneries, there was related growth in support industries. The principal support industry was the mass production of irrigation equipment for expanding orchards.[46]

However, generally speaking, the economy was suffering during the 1930s and this also was associated with certain ecological disruptions. The Great Depression resulted in the failure of many area businesses, and by the late 1930s many growers were uprooting fruit trees. By 1941, developers had leveled 4,500 acres of prune orchards, a trend that would become more common in the next decade.[47]

The post–World War II years saw the emergence of a new dimension of environmental concern with respect to the canneries' impact on water sources: pollution of the San Francisco Bay and Santa Clara Valley watersheds. A 1946 report on this problem concluded:

> The disposal of raw or inadequately treated sewage and industrial wastes in the waters of the Lower San Francisco Bay has destroyed its esthetic character. At times and seasons it has created a noisome mass, evil to look upon and disagreeable to smell. Fish life has been largely destroyed in many of the sloughs and in the lower (southerly) reaches of the Bay. Recreation is no longer pleasurable or healthful in these parts. Commercial developments, such as the Port of San Jose, for example, must await the advent of better conditions before they can be realized effectively.[48]

This report went on to call for "a complete about face" with regard to industrial pollution practices in order to save the local watershed. The document sums up nicely the contemporary EJ movement's definition of "the environment"—where we "live, work, and play"—in that the authors broadened their net beyond the watershed to include the impact of pollution on neighborhoods, the economy, and recreation,[49] showing that human beings are intimately connected to and affected by every function of ecosystems.

It is also noteworthy that, like the canneries themselves, the water contamination was centered mainly in the northern part of the city, where many communities of working-class and immigrant people and people of color were located. While this proximity may have been the result of convenience, it also represented a degradation of the local community's environmental integrity—an environmental injustice *outside* the workplace. Furthermore, one of the solutions to this problem of contaminated sewage was to propose the construction of a sewage treatment plant in Alviso, the northern part of the metropolitan area that was previously farmed and inhabited by Chinese and Japanese immigrants (many of whom remained) and was now also occupied by many Chicanos and Mexicanos.[50] This plant was eventually built and billed as the Bay Area's "most advanced sewage treatment" facility.[51] However, in 1997, the California Public Interest Research Group (CALPIRG) gave the plant a failing grade for violating the Clean Water Act and for dumping heavy metals and other toxics into the Bay.[52]

In March 1946, the State Board of Public Health adopted a "Resolution against Disposal of Raw Sewage into the Waters of the State without Appropriate Sewage Treatment." Furthermore, the California State Fish and Game Commission issued a directive requiring all food processors to install special screens in their plants to filter out and collect fine solid wastes that might otherwise pollute water sources. The amount of untreated sewage contaminating Bay Area water was becoming a problem and the citizenry and area governments finally had to take notice. The problem of polluted or toxic sewage would haunt the canning industry from the 1940s through the 1970s, and, soon after, would plague the electronics firms of Silicon Valley.

The County of Santa Clara opened the New Year in 1948 with an acute water shortage and an inadequate sewage treatment system. Five years later, the City of San Jose was cited by state officials for continuing to allow sewage dumping into the San Francisco Bay.[53]

During the 1960s, federal environmental laws began to have a major impact on the cannery industry and may have been the final blow to this ailing sector of Santa Clara County.[54] In 1975 and 1976, cannery sewage taxes increased 250 percent and cannery representatives were predicting a total collapse of the industry.[55] But there was a documented need for pollution control in the area. As we demonstrated earlier, Bay Area water sources had been compromised by the canneries since the 1940s, and the 1970s were no different. Before the fee hike, a cannery could dump 365,000,000 gallons of sewage each year and pay only $51,000.[56] The fee hike would double this bill, a price many environmentalists might argue pales by comparison to the ecological havoc wreaked on the Bay as a result of the dumping.[57]

The Canning Industry Declines

Within six years the county had lost seven more canneries. By the early to mid-1980s, the cannery industry was all but a historical artifact and most visitors to the Bay Area only associate the word "cannery" with the tourist attraction in Monterey, California, known as Cannery Row.

The canning industry reached its peak in 1954 with a fruit crop valued at more than $40 million and began to decline shortly thereafter. The reasons for this decline are many, including the emergence of California's Central Valley (to the south of Santa Clara) as the new premier agricultural region, continued problems with labor militancy among cannery workers, and the allegedly high cost of complying with environmental standards.[58] As of 1977 the industry still employed 12,000 workers, but this was down from an estimated 30,000 during the 1940s and 1950s.[59]

Despite all the turmoil and conflict, Santa Clara County has always held an image of itself as unique and immune to the economic volatility that periodically befalls the rest of the nation. One year into the Great Depression, for example, the local newspaper, politicians, and business leaders were proclaiming that the county's industries would not be affected by the economic downturn (a proclamation that would be proven false).[60] This arrogance would be repeated during the recessions of the 1980s, when many electronics firms could no longer honor the "no layoff pledges" they had made under the prior assumption that electronics was an entirely unique sector, immune to economic volatility. Another component of this superiority complex was the business community's self-serving position on welfare. During the initial months of the Great Depression, one industry leader remarked, "Government charity is, in its moral effects, a form of slavery."[61] Yet, within a few years, the prune industry was more than happy to receive a massive bailout from the federal government in the form of continuous aid that rescued that sector from disaster. From 1933 to 1937, the U.S. government purchased nearly 97 million pounds of prunes and offered generous loans to the industry through the Commodity Credit Corporation.[62] The same dynamic would play itself out during the post–World War II years, as the electronics industry received billions of federal dollars to develop itself, all the while claiming that genius and hard work were the driving forces behind the Silicon Valley "miracle." In this way, Santa Clara County businesses were never prepared for the boom-and-bust cycles that would periodically rock the local economy, yet they were almost always able to count on a public bailout: corporate welfare.

Finally, as difficult as the conditions were for cannery workers in Santa Clara County, their unions provided basic advantages that non-unionized workplaces generally did not offer. These imperfect but invaluable organizations would be the last that the Valley would see for a long time. The electronics industry of Silicon Valley not only displaced the canneries, it also destroyed all union activity and maintained a nearly perfect record of defeating organizing drives throughout the second half of the twentieth century.

4

The Emergence of Silicon Valley

High-Tech Development and Ecocide, 1950–2001

The Military-Industrial-University Complex

The high-tech electronics industry is generally believed to be the solution to economic development, environmental protection, and social equity needs around the globe. Experience, however, indicates that this industry is linked to continued and rising rates of poverty and economic volatility, ecological devastation, and social inequality around the world.

International Business Machines (IBM) built its first West Coast plant, its Pacific headquarters, in San Jose in 1943. The Food Machinery and Chemical Corporation (FMC), a locally based manufacturer of equipment and chemical sprays and fertilizers for canneries and farms, expanded its operations in San Jose to include the building of armaments for the World War II effort in the early 1940s.[1] The location and expansion of these and other major defense industry manufacturers signaled that San Jose was officially out of the Great Depression and beginning its transition into a new, post-agricultural economy. Predictably, Anglos—many of them Dust Bowl migrants—were the primary beneficiaries of these higher-skilled, higher-wage jobs in San Jose.[2] This meant that European immigrants and Chicanos/Mexicanos continued to hold down jobs in the lower-wage cannery sector, which would soon be in decline.[3] The electronics sector was emerging as the next economic stage in the Santa Clara Valley's evolution.

Before World War II, the electronics industry was concentrated in the Midwest and Northeast, producing mostly vacuum tubes and other products, generally for consumer radio markets. About 75 percent of all products were for consumer use. But by the end of the Korean War, the military accounted for 60 percent of the electronic products market.[4]

Facilitating this shift were the development of transistors (replacing the vacuum tube) and the virtual merger of the aerospace and electronics industries after the late 1950s. There was an ever-increasing demand for electronics products.

Stanford University was at the core of the transformation of Santa Clara Valley into Silicon Valley. Leland Stanford, the industrialist who endowed and founded the university, made a fortune supplying mining equipment during the Gold Rush, cofounded the Central Pacific Railroad, and advocated open Chinese immigration to the United States for railroad construction.[5] Frederick Terman, a professor of radio engineering at Stanford and later dean of engineering and provost of the university, was intent on institutionalizing the relationship between the university and the government. He viewed such a relationship as the future of U.S. research universities, and perhaps the only way to ensure U.S. military security in the world. In his vision, university-government alliances for military research would bring "better health, better transportation, more leisure, and a higher standard of living" to his faculty and to the general citizenry of Palo Alto, California, where Stanford was based.[6]

When the cold war began in the late 1940s, the U.S. Department of Defense was making large grants to several firms and universities (like Stanford) in Santa Clara County to create a center for military and defense industry research around the transistor and microprocessor. Stanford University was a national leader in research and development in this field. The Stanford Research Institute (SRI) was founded in 1946 by several Bay Area financiers and University officials for the purpose of conducting applied research for California's industrial and defense sectors. The Stanford Industrial Park (SIP) was founded in 1951 when the university leased land to electronics firms, including future giants like Hewlett-Packard and Varian Associates, who were among its first tenants. At SIP and nearby, many firms located and/or expanded, including FMC, Lockheed, Philco, General Electric, Sylvania, Fairchild, Memorex, National Semiconductor, and dozens of others. In return for prime business locations at the SIP and lucrative federal contracts, firms would endow department chairs, provide needed buildings and equipment for university laboratories, and sponsor relevant projects.[7] For their part, municipalities would provide tax relief and other subsidies and clear tracts of land for industrial development. Thus, Silicon Valley was the creation of the federal government, local municipalities, universities, industry, and the military. These stakeholders collaborated to produce an

TABLE 4.1

*Top Twelve Defense Contractors in Santa Clara County, 1988**

Contractor	Defense Contracts (in millions of $)
Lockheed	1,910
Ford	295
Westinghouse	282
Autek Systems	126
Stanford Telecom	88
Unisys Corporation	87
Varian Associates	51
TRW	35
Honeywell	29
Applied Technology	27
Teledyne	24
Hewlett Packard	21

*Smith and Woodward 1992, 4.

economic engine that would lead the nation—and indeed the world—in the production of electronics and computer chips for commercial, military, and consumer use.

The cold war economy provided Santa Clara County's defense/electronics industry with an opportunity to reap a windfall of government contracts. Since 1940, the infusion of federal money—hundreds of billions of dollars—has underwritten the expansion of high-technology industry.[8] By the late 1970s, Santa Clara County was receiving $2 billion annually in federal defense contracts, and the trend continues today (see Table 4.1).[9] Given this enormous volume of federal contracts pouring into the Valley, the claims that this industry started "from scratch" or that "brainpower is the most important factor in the semiconductor industry" defy reality.[10] Federal dollars went toward the development and production of "electronic warfare technology": the Trident missile, military lasers, rocket booster motors for the Air Force Titan 3C, spy satellites, and armored personnel carriers.[11] By 1979, two hundred thousand persons in the county were directly or indirectly employed in the electronics industry.[12] Thus, military demands shaped technological and economic developments in Santa Clara County from the beginning of the electronics industry.[13]

In 1955, William Shockley, co-inventor of the transistor, arrived in Palo Alto to found a company that would produce semiconductors. Shockley would in many ways spur the development of Silicon Valley, through the growth of a family tree of name-brand firms over the next several decades. Shockley Transistor Corporation quickly picked up several federal military contracts. But in 1957, all eight of Shockley's engineers defected and went to work for the newly founded semiconductor division of Fairchild Camera and Instrument Corporation (the Fairchild family was a cofounder and one of the principal stockholders of IBM). In 1961, Fairchild struck gold when it developed the ability to mass-produce integrated circuitry on fingernail-sized silicon chips. This process led directly to the production of the second generation of Minuteman intercontinental ballistic missiles (ICBMs) for the U.S. Air Force.[14] Charles Sporck, general manager at Fairchild, left to work at National Semiconductor in 1967, and Robert Noyce, one of the original "traitorous eight" engineers who defected from Shockley, resigned as head of Fairchild to cofound Intel, the largest producer of semiconductor components in Santa Clara County. Fairchild, National, and Intel became known as "The Big Three" in the industry. The development of large-scale integrated circuitry (leading to the boom in hand-held calculators) and the microprocessor—a programmable minicomputer on a chip, which led to the explosion of consumer use of computers—solidified Santa Clara County's position as the electronics industry leader during the 1960s and afterward. The result was an explosion in the number of firms—800 electronics businesses sprang up between 1950 and 1974. As hundreds of firms located, expanded, or were created in Santa Clara County, thousands of job seekers, many of them immigrants and people of color, came looking for work.

Immigrants and People of Color Seek Opportunities in the Valley of Dreams

Long an elite and exclusive private institution, Stanford University had no less stringent expectations of the firms it was allowing to locate in its industrial park. The university sought to ensure that the companies and their personnel would "blend in" with the suburban environment. They sought

> "light industry of a non-nuisance type . . . which will create a demand for technical employees of a high salary class that will be in a financial position to live in this area." The well-paid, well-educated worker in the industrial park made a "very desirable kind of resident" for the commu-

nity, according to the president of the Stanford Board of Trustees; employers should expect the suburban environs of the park to "attract a better class of workers."[15]

The terms "desirable" and "better class" are classic *code words* that denote Stanford's wish to attract white, middle-class workers and to repel undesirables, the multiracial blue-collar "riffraff" generally associated with manufacturing industries.[16] This was a textbook pitch reminiscent of the turn-of-the-century "city beautiful" efforts by Frederick Law Olmstead and others, who were openly classist and racist in their consultations with cities regarding what a desirable citizen population might be.[17] The *San Jose News* did its part to promote Santa Clara County's clean and lily-white image when it boasted that political and business leaders were able to attract entrepreneurs to the area who used "[i]mproved production techniques and new types of industries which eliminated or reduced industrial nuisances."[18] In 1968, while the Black Panther Party, the Brown Berets, La Raza Unida, and the San Francisco State University strike were in full force, the *San Jose News* proudly proclaimed that Santa Clara "is a white collar county . . . and 93.6 per cent of the county population [is] white."[19] Little did the Stanford University administrators, newspaper editors, and urban planners in the county realize that the region would soon be transformed into one of the most environmentally polluted and racially and ethnically diverse in the nation.

Chicanos

The 1960s and 1970s saw the emergence of public protests in San Jose's Chicano community when activists and residents put the spotlight on police brutality, economic discrimination, and the lack of services in their neighborhoods. Punctuating some of these concerns, one Chicano activist, Sal Alvarez, told an audience at a West Valley College symposium in 1969 that "Mexicans in San Jose got cheated out of land under urban renewal . . . [and] the killing of Mexican Americans by police in California is a very serious problem and a worry in San Jose."[20] Community-based and national organizations like United People Arriba, the Council on Latin American Affairs, and La Raza Unida were active in San Jose and Santa Clara County. Apartment complexes, construction companies, schools, and police departments charged with discrimination against Chicanos were the primary targets of the protest movement.[21]

A 1973 Rand Corporation study reported that while much of San Jose was experiencing prosperity, "poor [largely Latino] neighborhoods [had] deteriorated relative to better-off neighborhoods and segregation had increased."[22] Housing quickly became a social and environmental justice issue as activists linked poor housing quality and poor health with racism. The fate that befell so many African American neighborhoods around the nation during the 1950s due to urban renewal efforts also came to pass in the barrio of Sal si Puedes in east San Jose.[23] Urban renewal was the federal effort to improve municipalities by razing degraded and blighted housing and by connecting cities and states through the interstate highway system. These changes devastated and destroyed many communities, earning urban renewal the nicknames "Negro removal" and "Mexican removal." For example, most of the houses of Sal si Puedes were razed as part of an urban renewal program that included the construction of a new freeway.[24] In the 1980s, other neighborhoods populated by people of color in San Jose would fall prey to urban renewal—ironically, for the construction of the Tech Museum, the city's monument to the electronics and computer industries, and its own physical proclamation of its status as the "Capitol of Silicon Valley."[25]

The Chicano community of Alviso, in North San Jose, was also the site of many conflicts over environmental and social justice concerns. In March 1973, activists set up a tollbooth on the Gold Street Bridge, charging twenty-five cents for cars to pass—a symbolic protest to call attention to the lack of paved streets in Alviso. The streets were not paved until 1980, twelve years after Alviso was annexed by San Jose.[26]

Asian/Pacific Islander Americans and Immigrants

According to the 1980 census, an estimated 11,700 Vietnamese immigrants were living in San Jose. By 1987 that number had jumped to around 75,000.[27] During the 1980s, there was a marked rise in white resentment and hate crimes directed at the Vietnamese population in particular, and at Asian/Pacific Islander Americans and immigrants in Silicon Valley in general.[28] During this period, many manufacturing workers were being laid off in the United States and there was increased tension over the heightened competition with Japan in the auto and electronics industries. Much of this tension was channeled into scapegoating any and all Asians, regardless of their specific ancestry.[29]

At the same time, some Asian immigrants were doing well in the high-technology sector, as white-collar workers. Chinese software engineers set up many businesses by networking with firms and entrepreneurs overseas and, by 1992, Asians made up a third of the engineering workforce in Silicon Valley.[30] Very few of these upwardly mobile Asian workers broke through the glass ceiling into the ranks of management. David Lam, founder of Lam Research, a company that makes equipment used in chip making, explained, "Many Asian engineers are not being looked at as having management talent. They are looked upon as good work horses, and not race horses."[31]

As financially stable as some Asians appeared to be, most were working-class and many of them experienced plenty of setbacks. In 1983, Atari Corporation, the video game maker, announced that it was closing its Silicon Valley plant and shifting production overseas to Hong Kong and Taiwan. Most of those production employees thrown out of work were Latino, Chinese, and Vietnamese. One of these workers was Hoa Ly. In 1978, Ly escaped Saigon in a fishing boat jammed with fifty-five other refugees. He spent half a year in a Malaysian refugee camp and later moved to Silicon Valley, where he got a job in printed circuit board assembly at Atari for which he was paid $7 per hour. Since his mother and father both worked by his side at Atari, when the company closed down his whole family was suddenly out of work. The Atari case was also quite significant because it was a high-profile example of the empty "no layoff pledge" many Silicon Valley companies made during the 1970s, when business and political leaders were claiming that this industry was immune to recession.[32]

Since the 1980s, the popular sentiment regarding immigrants took at least two, perhaps contradictory, directions. The first was a continuation of the traditional nativist approach to immigration, and this anti-immigrant movement gained sanction at the highest levels, as the efforts to pass Proposition 187 (denying undocumented residents access to public services) and the implementation of the Welfare and Immigration Reform Acts of 1996 moved forward. Television commercials depicted gangs of "illegals" crossing the Rio Grande under cover of night, and presidential hopeful Pat Buchanan warned of a "foreign invasion" from Mexico. In Silicon Valley, as elsewhere, this backlash took the form of

Immigration and Naturalization Service (INS) raids of workplaces. In April 1984, when the INS opened a new office in Silicon Valley, they conducted raids on two electronics firms in San Jose and the city of Santa Clara on the same day. Using classic nativist rhetoric, Harold Ezell, the INS's Western regional commissioner, stated, "Probably 25 percent of the working population in this area is here illegally, particularly in the Silicon Valley area. We intend to make our presence known. Our officers will be *freeing up jobs for U.S. citizens* and people who are here legally."[33] Neither the newspaper article nor the INS acknowledged that, without the abundance of undocumented immigrant labor, many Silicon Valley industries would relocate overseas or south of the U.S./Mexico border. The second direction in popular sentiment was the general consensus among labor, business, and politicians that the Immigration bill of 1990 and the H1-B visa expansion in 2000 (allowing white-collar high tech workers to immigrate to work at a particular firm for a limited period) were good for the U.S. economy, because of the alleged "shortage" of skilled workers.[34] The AFL-CIO called for an amnesty for the nation's estimated 5 million undocumented immigrants and Federal Reserve Board Chairman Alan Greenspan warned that, without a new wave of skilled immigrant workers, the integrity of the U.S. economy would be threatened.[35] Soon the national and California state legislation and policy making around these questions lumped documented and undocumented immigrants together for punitive measures. This indiscriminate grouping of these populations rendered the distinctions moot in many cases and revealed the deep similarities between the racist nature of immigration policy at the turn of the twentieth century and the twenty-first.[36] One of the major differences for Silicon Valley and the state of California, however, was that, for the first time since indigenous peoples occupied the land, people of color were now the majority. Of the 1.6 million people in Santa Clara County in the year 2000, 49 percent were white, 24 percent were Latino, 23 percent were Asian, and 4 percent were African American.[37]

Immigrants, People of Color, and Inequalities in the New Economy

The level of social inequality in Silicon Valley is a byproduct of an unregulated quest for wealth. In a 1979 special report on the state of Silicon

TABLE 4.2

*High-Technology Employment by Race, Ethnicity, and Gender in Santa Clara County, 1970**

	Male	Female	White	Latino	Asian
Managers	85%	15%	88%	4%	5%
Professionals	82%	19%	83%	3%	12%
Craft Workers	56%	44%	63%	17%	14%
Operatives	31%	69%	49%	23%	19%
Laborers	38%	61%	41%	34%	17%

*Adapted from Equal Employment Opportunity Commission 1980.

Valley, the *San Jose Mercury News* noted that "[t]he economic gap between the white majority and the minorities and poor is widening, and inflation will add to the inequities, putting greater strain on the social system." That same report stated that, although most of the new immigrants in the area have either a college or a trade school education, the majority are working at jobs "generally below their skill level . . . [and] more than half of them are employed in the valley's electronics industry."

By 1995, the situation had not improved. On the one hand, racial diversity was increasing: the total Latino population constituted 23 percent of the county's residents, Asian/Pacific Islanders were 20 percent, and the white population was less than 53 percent.[38] However, Silicon Valley was experiencing ever-increasing wage inequality across race, class, and gender, at rates much greater than in the United States as a whole.[39] Perhaps more disturbing than this growing inequality was an absolute and relative decline in wages for many workers.[40]

Carrying on the historical tradition of gender and racial segregation in the Santa Clara County workplace, Silicon Valley electronics firms surpassed the national economy in their ability to divide employees by social categories (see Tables 4.2, 4.3, and 4.4). As one study reported, "The workforce is sharply divided along ethnic, sexual, and educational lines. In general, white males hold the positions with the highest incomes and greatest power. Non-white men (including Hispanics) and white women fall in the middle of the high tech hierarchy. And minority women stand at the bottom of the occupational structure."[41]

Table 4.2 reveals that, in 1970, women and people of color constituted an exceedingly small fraction of professional and managerial positions in the Valley; these same groups were concentrated in the

TABLE 4.3

*High-Technology Employment by Race and Ethnicity in Santa Clara County, 1980**

	White	Latino	Asian
Population	58%	20%	17%
Managers	80%	4%	10%
Professionals	70%	3%	20%
Craft Workers	45%	20%	30%
Operatives	25%	23%	47%
Laborers	19%	36%	42%

*Adapted from Siegel 1994. The above percentages are rough approximations based on a reading of bar graphs.

lower-paid, higher-risk occupations of craft worker, operative, and laborer. Ten years later, these patterns remained virtually unchanged, except for the class bifurcation among the Asian populations, as many skilled Chinese immigrants moved into higher-paid positions, while the Vietnamese population was concentrated in the lowest-paid positions (see Table 4.3). Thus race, class, and gender operated in ways that generally disadvantaged people of color and women in Silicon Valley. The problem of racial and gender discrimination in the industry was so serious that the U.S. Commission on Civil Rights held a hearing on the topic in San Jose in 1982.[42]

In addition to a racial and gendered division of labor in Silicon Valley,[43] residential segregation was rampant in the region. Wealthy, educated whites tended to cluster in the Palo Alto area and the foothills, while less affluent residents, including most people of color in the Valley, concentrated in the "flatlands" of East Palo Alto, Mountain View, and San Jose.[44] This spatial segregation by race and class also produced and reinforced school segregation.[45]

The Underside of Silicon Valley: Land Grabs, Toxics, and Ecocide

Land Annexation and Sewage Troubles

Paralleling the Spanish Conquest, the Gold Rush, and the cannery and orchard economies in the Valley, San Jose's industrial and political leaders in the post–World War II era sought to expand their profits and their influence. As two political scientists flatly put it, "Their policy was *growth*,"[46] and pollution was one of the many negative byproducts of this approach to development. Under City Manager Dutch Hamman, San Jose allowed "developers [to] do what they wanted wherever they wanted to do it."[47] The city was intent on achieving growth, attracting industry, and increasing the tax base at almost any price:

> Simply put, growth meant prosperity, and bigger was better. Few people identified growth as anything but beneficial for a community, and the thought of the contrary was virtually unthinkable. Growth meant progress, and progress was the road to prosperity. In this atmosphere it is not difficult to understand why proposals for slower and more coordinated growth were neglected.[48]

Soon San Jose was caught up in what became known as the "annexation wars." The city coerced and cheated farmers, residents, and others out of large tracts of land all over the county.[49] This was a contemporary extension of the "land grabs" that had been ongoing in the area for nearly two centuries.

The annexation wars took on an interesting environmental twist during the 1940s and 1950s. At this time, surrounding towns needed sewage treatment facilities, and San Jose, in a much despised political move, gave them a choice: annexation as the price for sewage service or no sewage service at all. San Jose had built an extensive "outfall" sewage drainage system into the San Francisco Bay in the 1880s. This was the first and largest sewage system in the South Bay and was built to accommodate the significant effluent produced by the canneries and any waste from future (projected) population growth. So other cities either had to

TABLE 4.4

*High-Technology Employment by Race, Ethnicity, and Gender in Santa Clara County, 1990**

	Male	Female	White	Latino	Asian
Officials/Mgrs	78%	22%	80%	4%	12%
Professionals	72%	28%	70%	4%	21%
Technicians	76%	24%	53%	10%	31%
Office Workers	21%	79%	65%	13%	15%
Blue-Collar	61%	39%	43%	20%	31%
Laborers	51%	49%	18%	38%	42%

*Siegel 1994.

build their own systems or become a part of this growing giant in order to handle their own waste and comply with state and federal laws. This was San Jose's way of using its monopoly on sewage toward its goal of growth at all costs. As Dutch Hamman bluntly stated, "We're in this fight to the finish, and if we have to use sewage disposal to bring Santa Clara [County municipalities] to some point of reasoning, we'll do it."[50] The annexations continued; between 1950 and 1960 the city added 491 lots, and it added at least 900 more between 1960 and 1970—for a total annexation of 132 square miles of land.[51] This pattern continues through the present day, underscoring that the most basic natural resources—land and water—remain critical for the survival of the high-technology industry, as was the case for each major system of production that preceded it.[52]

During the 1970s, San Jose's sewage treatment plant was operating near, at, or over peak capacity and was increasingly toxic. A sewage spill from the plant in 1979 killed off a great deal of marine life in the southern part of the San Francisco Bay. Another spill just one year later exacerbated the situation. This sewage plant was located in the Latino community of Alviso. By this time, the sewage problem was attributable less to cannery pollution than to electronics industry pollution. Industry representatives from National Semiconductor complained that, by paying sewage fees and taxes, the electronics sector was being forced to "subsidize the canneries"[53]; in actual fact, however, National Semiconductor was one of the biggest polluters in the Valley and has since been sued for allowing its workers to be poisoned and afflicted with cancer. More generally, the high-tech industry is responsible for dumping sewage laced with heavy metals like nickel, cyanide, lead, and cadmium into the Bay, contaminating the oyster and shellfish population to the point that they are unsafe for human consumption today.[54] A 1988 study by the environmental group Communities for a Better Environment found that nine of the top twelve releasers of toxic heavy metals into the local ecosystem were high-tech firms. An investigation by the *San Francisco Bay Guardian* newspaper found that tech companies were regularly dumping toxics into the public sewer system, violating local and federal clean water laws, and contaminating the Bay waters just as the canneries and gold miners had done in previous generations.[55]

Pushing against the dominant political grain, a number of growth control advocates were operating in the area, although they wielded significantly less power than the pro-growth forces. They included farmers, certain school districts, and conservationists who sought to achieve a balance between the aesthetic agricultural side of the Valley and the growing industrial areas. Farmers understandably felt threatened by San Jose's unchecked growth, so many of them fought the annexations. In 1953, a task force of agricultural leaders pressured the county to develop a new zoning classification called "exclusive agricultural" areas, also known as the "green belt" or "green zones."[56] In 1955, farmers succeeded in obtaining passage of the Agricultural Exclusion Act as well. This act mandated that land zoned for agriculture could not be annexed to a municipality without each property owner's consent. The 1957 Agricultural Assessment Law and the more recent Williamson Act both allowed for voluntary contracts between landowners and the city and county governments to maintain land for agricultural use while taxing properties based on their agricultural—not real estate market—value.[57] By 1958, 40,000 acres fell into this zoning category, and by 1960 that number had increased to 70,000 acres.[58]

From "Clean Industry" to the Valley of Toxic Fright

As others have noted, the new economy, based on the emerging electronics industry and its defense and warfare orientation, was ushered in with a fanfare that promised new jobs in a "clean industry."[59] Stanford University's Industrial Park also bought into (and propagated) this myth:

> Stanford Industrial Park represented a new variation on several traditions, and observers initially were unsure how to describe it. Some likened the district to a garden, others to a park, still others to a country club—all provocative comparisons for what was at bottom an industrial workplace. One booster simply listed the amenities—"broad lawns, employee patios, trees, flowers and shrubs, walls of glass, recreational club"—that made it such a "pleasant place" in contrast to "smokestacks, noise, coal, cars, soot, and other things" associated with industry in the East and Midwest. All these features defined the park as a western hybrid of the two mid-twentieth century strains of suburb and campus. . . . Stanford consequently limited buildings in the park to no more than two stories, forbade smokestacks, and prohibited any noises, odors, or emissions that might offend homeowners.[60]

This postindustrial, post-smokestack, campus-like suburban planning made it easy for developers and industry owners to claim that the electronics sector was "clean" and "pollution free."[61] The clean image of the electronics industry was touted by executives, politicians, and newspapers everywhere. Harold Singer, an official of the San Francisco Bay Regional Water Quality Control Board, once stated, "the horizon above San Jose is unmarred by smokestacks, and people here are proud of that. They have worked hard at making the valley a base of the computer-electronics industry and an unpolluted place to live."[62]

As recently as the year 2000, the *Smithsonian Magazine* described the "clean rooms" where microchips are made as "the most fanatically clean, most thoroughly sanitized places on the planet," where "one could eat one's oatmeal off the floor."[63] The highly toxic wafers from which microchips are cut are viewed by industry promoters as "pristine,"[64] and the chemical-laden water that washes semiconductor components in the electronics "fab" plants is described as "pure."[65] Even former U.S. President Bill Clinton rubbed shoulders with CEOs in Silicon Valley in the 1990s, publicly proclaiming that the high-tech industry "will move America forward to a stronger economy, a cleaner environment and technological leadership."[66] These accounts leave the uninformed reader with the impression that high-tech firms are the paragon of hygiene and safety, sanitation and environmental responsibility.

The history behind this image—and more importantly, the need and desire to create this image—seems to be lost on most observers of Silicon Valley's environmental problems. In the 1940s, as large defense industries such as General Electric, IBM, and Westinghouse Electric were locating and/or expanding in the Valley, the agricultural interests were quite concerned about the environmental impacts of this new form of development. In 1950, the San Jose Chamber of Commerce wrote:

> there were some sincere and intelligent people who looked askance at this industrial development. They had genuine fears that smokestacks would "encircle the city"; that "blighted areas" would spring up in industrial sections; that orchards would be torn up "by the hundreds"; and that by past standards, this accelerated trend in the establishment of a new industry might result in an unbalanced, top-heavy economy destined to collapse at some undetermined time in the future.[67]

The author of a study of the Bay Area conducted during the 1950s described the level of accuracy of these dire predictions:

> The fears that orchards would be torn up by the hundreds were indeed well founded; and smokestacks, though they by no means encircled the city, did undeniably contribute to the development of a new problem. The Chamber [of Commerce] sought to assure the skeptical that industrial growth was not incompatible with desirable living conditions. Yet it was not long before the Santa Clara County Board of Supervisors found it necessary to designate the entire county an air pollution control district; the skies over the Santa Clara Valley were becoming a dirty gray. In March, 1950, the county health officer, declared that "smog," the murky atmospheric condition familiar to Los Angeles, was not only a Santa Clara County problem but also a "Bay Area problem." The new factories in San Jose and Santa Clara were producing air pollutants. [68]

Thus the very beginnings of the electronics and defense industries were marred by environmental problems, the greatest irony of all being that they involved "smokestacks." So it seems that in the 1950s and 1960s, when Palo Alto and other cities in the Valley demanded that new industries be smokestack-less, these concerns were rooted in the real experience of having observed this sector befoul the local environment. Whether or not the municipalities were aware of it, these expectations and claims of nonpollution were either naïve promises or lies, while the production processes internal to these corporations actually became even more toxic than before. This heavily polluted past also challenges earlier claims by Silicon Valley boosters that the region was "pollution free" prior to the electronics boom of the 1960s.[69]

The End of Innocence

In the 1960s and 1970s, a sort of amnesia fell over the Valley's residents and policy makers; they all seemed to arrive at a consensus that the new industry was somehow pristine. This image was shattered in December 1981, when it was discovered that the drinking water well that supplied 16,500 homes in the Los Paseos neighborhood of South San Jose was contaminated with the deadly chemical trichloroethane (TCA), a solvent used to remove grease from microchips and printed circuit boards after they are manufactured. Officials estimated that 14,000 gallons of

TCA and another 44,000 gallons of various toxic waste materials had been leaking from an underground storage tank for at least a year and a half.[70] The responsible party was the Fairchild Semiconductor corporation.

Lorraine Ross, a resident of the neighborhood (and a mother), was catapulted into the role of environmental activist. She and her neighbors mapped out a disturbing and pervasive cluster of cancers, miscarriages, birth defects, infant heart problems, and fatalities in the neighborhood that public health authorities and the industry were forced to take seriously. Two health studies were carried out immediately, both of which confirmed the presence of higher than expected frequencies of congenital birth defects (three times the normal rate), spontaneous abortions, and heart defects. However, neither study would take the bold step of pinpointing industrial chemicals as the cause.[71] One reason for the delayed discovery of the presence of chemicals was that at the time, state and federal regulations did not require testing for industrial chemicals. Tests were only required for viruses, bacteria, pesticides, and herbicides.[72] A painful irony here is that TCA was commonly used as a substitute cleaner for TCE (trichloroethylene). TCE is a suspected carcinogen that was nearly phased out of the industry in the late 1970s—the result of the Campaign to Ban TCE, led by occupational health advocates in San Jose.[73] This community organizing success was reversed when the industry phased out the targeted chemical and substituted a comparably hazardous one.

Back in the Los Paseos community of San Jose, Lorraine Ross was organizing against Fairchild, berating the city council that year with the question, "Fairchild or my child?": "It takes a lot of nerve for them to invade a pre-existing residential neighborhood, pour dangerous chemicals into a leaking tank, poison the surrounding environment and hide the fact from the people affected by their negligence."[74]

Ross was not alone. She was a leader of a burgeoning antitoxics/environmental justice and occupational health movement taking shape in Silicon Valley. Organizations like the Santa Clara Center for Occupational Safety and Health (SCCOSH) had prior to the contamination of Los Paseos been involved in leading community workshops on chemical solvents such as those spilled at Fairchild.[75] The Silicon Valley Toxics Coalition (SVTC), an environmental justice group that had formed in response to the Fairchild spill, was also at the forefront of the campaign to bring that company to justice. In 1983, Fairchild closed down its plant in South San Jose, a victory for Ross and the environmental justice movement. Since then, the company has spent more than $40 million on the cleanup. Similar chemical accidents and resulting toxic illness and death tolls have occurred in the communities near the IBM and Teledyne Semiconductor plants; the IBM spill is one of the largest in the county, having leaked toxics since 1956.[76] IBM had to install a more extensive chemical detection system for ground water monitoring as a result of efforts by the Silicon Valley Toxics Coalition, Communities for a Better Environment (a San Francisco–based group), the Santa Clara Valley Water District, and the County Board of Supervisors.[77] This result actually set two new precedents. The first was that the so-called "clean industry" could no longer be viewed as pristine. The second was that this industry was not trustworthy and that only the presence of a strong local environmental justice movement could ensure that necessary reforms would materialize.

Challenges to the "clean industry" image in the Valley abound. The USEPA has estimated that a large area of land, contaminated by eleven electronics plants in the Mountain View area alone (a community with a large population of immigrants, people of color, and working-class persons) will take $60 million and 300 years to clean up. As we mentioned in chapter 2, that site is located at Moffett Field, the old Naval base that sat on the land owned by one of the few remaining Ohlones in the area, Lope Inigo. As for the once pure water and fertile land of the Valley, 57 private and 47 public drinking wells were contaminated as of 1992, and 66 plots of land have been declared too toxic for human beings to walk on.[78]

Soon after the Fairchild spill made news headlines, various arms of the federal government also took on the charge of addressing this problem. In 1985, the Congressional Committee on Public Works and Transportation held a hearing on toxic concerns in the electronics industry in San Jose.[79] The USEPA also undertook its own study of the human health risks associated with industrial pollution in Santa Clara County. The agency released a preliminary draft of this "Integrated Environmental Management Project" in 1985, which concluded that pollution-related health risks in the area were "comparatively low."[80] Clearly elated, a spokesman for the Semiconductor Industry Association (SIA) informed the media that, based on the EPA's findings, industrial chemicals in Silicon Valley "do not pose a significant threat to human life."[81] Environmentalists and journalists immediately became suspicious and discovered that a scandal was afoot. According to one report, long before the study was released, the SIA had hired a high-ranking official of the Republican Party who lobbied the EPA into lowering risk estimates and excluding tests from the study that would indicate whether solvents caused chronic diseases, birth defects, and miscarriages.[82] This effort seems to have paid off, as it appears to have shaped the "findings" in the EPA's study. But it was a major source of embarrassment for the EPA when high-profile environmental groups and political leaders denounced the report as flawed and biased in favor of industry. The true significance of this event was that environmentalists learned that they had to remain vigilant and be wary of "scientific studies," because the USEPA and any other agency could be bought and paid for by industry to silence or "disprove" dissenting perspectives concerning the true costs of the Silicon Valley "miracle."

After testing other companies for toxins, county authorities found that sixty-five of seventy-nine (or 82 percent) had hazardous chemicals in the ground beneath their plants. Some of these included IBM, Intel, Hewlett-Packard, DEC, Tandem, Raytheon, NEC, AMD, Signetics, TRW, and many others. Today Santa Clara County is home to twenty-nine Superfund sites, more than any other county in the nation, and twenty-four of those sites are the result of pollution by electronics firms.

The earlier claims that computer/electronics was a "clean industry" rang painfully hollow, because, as the Silicon Valley Toxics Coalition's Ted Smith put it, this industry had "buried its smokestacks underground."[83] Other leaders stated similar concerns:

> Voicing the shock shared by cities that had assumed the electronics industry was nonpolluting, San Jose's mayor, Janet Gray Hayes, said, "I remember thinking about smokestacks in other industries. I didn't expect this problem in my own backyard."[84] She continued, "When I first became Mayor and we embarked on an economic development program, there was no doubt in my mind that this was a clean industry. We now know that we are definitely in the midst of a chemical revolution."'[85]

Why Is High Tech Toxic?

Mayor Hayes got it right: we *are* in the midst of a chemical revolution, with more chemicals in use today than ever before, and with very little knowledge of their toxicological effects. In 1999, on average, the production of an eight-inch silicon wafer required the following resources:

4,267 cubic feet of bulk gases, 3,787 gallons of waste water, 27 pounds of chemicals, 29 cubic feet of hazardous gases, 9 pounds of hazardous waste, and 3,023 gallons of deionized water.[86]

Why was so much toxic waste being produced? First, the semiconductor chips themselves had toxics built into them. This was because silicon, a natural conductor of electricity, would increase its conductivity exponentially when certain chemicals were applied to it.[87] Second, it is now well known that the U.S. military is the largest producer of toxic waste in the nation. So the fact that Silicon Valley was hardwired for military production is a major reason for the presence of an inordinate amount of toxins in these industries. Specifically, the use of toxics is required in many contracts with the military. As one report details:

> Many products made by electronics firms are destined for use by the military. Military production specifications are the driving force behind product design. Yet, these specifications are unrelated to social needs or public costs. For instance, military specifications for many electronics devices demand the use of CFC-113 as a solvent of choice. CFC-113 is one of the main destroyers of the Earth's protective ozone layer. The military's need for faster, radiation resistant semiconductors has also prompted the development of gallium arsenide–based chips, and this demand has placed an extra burden for safety on the electronics manufacturers, since gallium arsenide production techniques require far greater amounts of carcinogenic arsenic.[88]

Third, the electronics industry was taking advantage of breakthroughs in research and development in the petrochemical industry:

> the microelectronics industry is developing in a new age of synthetic chemicals. Since World War II, major technical innovations in the petrochemical industry have produced a wide range of synthetic chemicals for industrial production. Early work by the National Bureau of Standards and the American Petroleum Institute during the 1930s shifted the focus of hydrocarbon production from coal to petroleum. Combined with the heavy defense investments in chemical research during the war, this led to an explosion of new chemicals on the market following 1945. The U.S. production of synthetic chemicals increased from about 1 billion pounds in 1940 to 30 billion in 1950 and 300 billion in 1976. The rapid increase in the quantity and variety of new chemicals paralleled

> the development of the microelectronics industry. Unlike older industries that developed when resources were more limited and naturally occurring, the high-tech industry capitalized on new solvents such as ethylene, benzene, and styrene, complex halogenated hydrocarbons like trichloroethylene and methylene chloride, and various new ketones and resins.[89]

Finally, much of the water pollution associated with the high-tech industry in Silicon Valley stems from historical practices within the agricultural sector:

> It is estimated that there are about ten thousand well pipes in the Valley which extend from the surface to a depth of 30 to 150 feet into the ground. These were well pipes for agricultural uses on the farms. When the factories were built throughout Silicon Valley, most of these well pipes were simply buried. No one knows any longer where the majority of these well pipes are located. After careful searches through the records of the water authorities and other governmental agencies, about three thousand old well pipes were located. The unidentified pipes which remain puncture the clay strata and permit chemically contaminated ground water to seep into the underground water supplies, whereby toxic substances are distributed far and wide.[90]

Thus, the use of toxics in this industry (and their resultant ecological impacts) emerged from "advances" (and past mistakes) made in the military, chemical, and agricultural industries.

Aside from toxics spewing out of the drainpipes of Silicon Valley's industries, the layout of the region and the growth of the population presented other environmental challenges. Traffic congestion on the interstates in Santa Clara County is legendary, and the resultant air pollution produces a haze that covers the sky and obscures the view of the nearby hills at just a few miles range.[91] In 1965, Frederick Terman of Stanford University argued that the traffic and air pollution were "really a pretty small price to pay" for the wealth generated in the Valley.[92] One opinionated observer who analyzed this situation called Santa Clara County "a hideous, smog-covered, amoeboid sprawl of housing tracts, freeways and shopping centers . . . [created] needlessly and mindlessly."[93] Skyrocketing housing costs made this area even less desirable for many workers pursuing a high quality of life, and less attainable for most lower-middle- and working-class families.[94] Hence, Silicon Valley was rapidly turning into a metropolis with all the problems of other major urban areas rearing their heads, thus substantially lessening the region's unique appeal. If the social, economic, and environmental ills of Silicon Valley were a burden for the general population, they were even more acute for immigrant communities and communities of color.

Environmental Inequalities

When the nature and extent of high tech's toxicity became public knowledge, a decline in the location of such industries in white communities was complemented by a shift to lower-income communities and communities of color.[95] For example, after the discovery of toxic waste in the water tables in Palo Alto, residents organized and supported new regulations so strict that some companies moved out or decided against locating there, and shifted their toxic production to less restrictive communities (with higher percentages of working-class people and persons of color) such as Mountain View.[96] Similarly, during the early 1980s in Sonoma County (north of San Francisco), citizens opposed *all* high-tech development because of the newly discovered associated pollution threats. And in 1984, a Fremont-based group, Sensible Citizens Reacting Against Hazardous Materials (SCRAM), organized to block CTS Printex's attempts to locate a plant in that town.[97] This dynamic was a stark departure from the fights and bidding wars among municipalities in their efforts to attract the "clean industry" in previous decades. But this is also a pattern we see in myriad other "environmental protection" practices that impact communities of color across the United States and around the world. In other words, immigrants and people of color bear the cost of both environmental destruction (when industry extracts or pollutes natural resources) *and* environmental protection (when white, affluent communities discover that an industry is toxic and protect themselves by shifting the burden onto lower-income neighborhoods and communities of color).[98]

However, some polluting businesses that originally located in marginal neighborhoods were forced to reform because communities of color were also organizing. For example, Lorenz Barrel and Drum was located in a working-class and Latino neighborhood in south San Jose. For forty years this company treated the electronics industry's hazardous waste. The site was located a half block away from Mi Tierra, one of San Jose's first community gardens. A local neighborhood group, Students and Community Against Lorenz Pollution (SCALP), formed and pressured authorities to take action.[99] In 1986, federal authorities shut the company down, citing criminal violations. The soil and ground water on site were found to be contaminated with at least fourteen toxic chemical compounds.[100] The site was "remediated" for a $5.2 million price tag, but was covered with an asphalt cap—the "cleanup" method of choice in low-income communities and communities of color—rather than subjected to a true abatement and restoration operation as preferred under law.[101] Even so, more than 25,000 drums containing hazardous waste and 3,000 cubic yards of contaminated soil were removed. A year later, the company's owner was sentenced to two years in jail, assessed fines of $2.04 million, and ordered to spend up to $100,000 on health monitoring for current and former neighbors and employees. So while environmental racism placed these residents and workers at risk, they were able to achieve some modicum of justice through collective action.

While residents and the general public may not have been knowledgeable concerning the hazards associated with electronics production before the 1980s, ample evidence suggests that the industry was aware of the facts early on. As we mentioned in the beginning of this section, industry and government were cognizant of the air pollution associated with the electronics/defense industries as early as 1950. Years later, in 1976 (five years before the Fairchild spill was made public), a study submitted to the Santa Clara County Board of Supervisors disclosed that tons of "poisonous and explosive chemicals" were being illegally dumped in communities and into the sewer system throughout the region. These hazardous materials were from electronics firms and many of these dumps were located in communities of color and low-income neighborhoods inside the county and beyond.[102] So communities of color were being polluted well before the public outrage against the electronics industry made headlines in the early 1980s, and the industry was well aware of it.

After communities began demanding that toxic sites be placed on the EPA's Superfund list, evidence of another pattern of environmental inequality emerged. Table 4.5 reveals that many of these federally designated toxic Superfund sites are in communities of color and working-class neighborhoods. For example, Mountain View (where a high per-

TABLE 4.5

*Superfund Sites in Santa Clara County, California**

City	Number of Sites
Mountain View	10
Sunnyvale	6
Alviso	1 (entire community North of Hwy 237)
Santa Clara	5
San Jose	5
Palo Alto	2
Cupertino	2

*Working Partnerships USA 1998, 52.

centage of immigrants and people of color live) has the most Superfund sites in the county (ten); the increasingly multi-ethnic Sunnyvale municipality has the second highest number (with at least four of these sites in extremely close proximity to two mobile home communities and the San Miguel Child Development Center). The working-class Chicano community of Alviso, with just one such site, seems to have gotten off easy. However, that one site covers the entire Alviso community north of Highway 237. Much of this community is contaminated by asbestos; it is home to six garbage landfills; and now high-tech companies that build "servers" for computer and Internet networks are seeking to construct these large electrical and fuel-based facilities there as well. This project will lead to the addition of 82 diesel generators and an estimated 1.2 million gallons of fuel stored on site in the community.[103] In contrast, the two most affluent cities (with the highest percentage of white residents) in the Valley—Palo Alto and Cupertino—host only two Superfund sites each and now have the most restrictive industrial and residential zoning requirements in the county.[104]

In addition to soil and water pollution, the air pollution from area industries was considerable and was also distributed unevenly by race and class:

> Census figures and TRI [Toxic Release Inventory] air emissions data show clear evidence of environmental inequality in Santa Clara County as of 1990. . . . TRI facilities are concentrated in [census] tracts where households have low to moderate incomes . . . some tracts that have higher-than-average Hispanic populations contain some of the greatest concentrations of TRI facilities.[105]

TABLE 4.6

*Relationship between the Latino Population and the Presence of TRI Emissions in Santa Clara County Census Tracts, 1990**

% Latino	All Tracts	Tracts with TRI Emissions (expected frequency)
23.5–82.0	75	17 (10.25)
13.2–23.5	75	12 (10.25)
7.0–13.2	75	8 (10.25)
0–7.0	75	4 (10.25)

* Szasz and Meuser 2000, 609.

Table 4.6 divides Santa Clara County's census tracts into four equal groups, according to the percentage of their population that is Latino. These data reveal that those tracts with the highest percentage of Latinos are more than four times as likely to host a TRI facility as are tracts with the smallest percentage Latino residents. Conversely, the greater the percentage of white residents, the smaller the pollution burden. This body of evidence reveals a "clustering of low-income, mainly rental households, and of certain communities of color and children in neighborhoods in close proximity to the valley's toxic industrial belt, along Highway 101."[106] The future of environmental inequalities looks no better, considering the results of the San Jose City Council's proposed solution to the "energy crisis" in California: constructing sixteen new power plants, fifteen of which are to be located in communities in which people of color make up 67 percent or more of the population.[107]

Finally, most persons working to produce computers and semiconductor chips *inside* Silicon Valley's electronics firms are women, immigrants, and people of color. Compounding the pollution in their neighborhoods, these workers hold jobs that are more toxic than those found in any other basic industry (see chapters 5, 6, and 7).

A Sustainable and Just Future for Santa Clara County?

As with Santa Clara County's previous historical eras, opportunities to opt for less resource-intensive practices in the semiconductor industry were often available. For example, after the Vietnam War, many activists promoted the "Peace Dividend" and the idea of "military conversion," whereby these industries and government funds would be redirected toward the public interest in some way.[108] The Mid-Peninsula Conversion

Project in Santa Clara County was a leading proponent of reorienting electronics production toward nonmilitary purposes. Unfortunately, these arguments were lost in the cold-war race to build up military and nuclear arsenals in the effort to maintain the United States's position as the world's dominant superpower. The continued focus on military production was a betrayal for many activists, and was arguably a slap in the public's face because so many taxpayer dollars had been allocated to fund this industry, the social benefits of which were questionable.

A second example of an opportunity to engage in sustainable planning occurred at the dawn of the electronics industry's growth spurt, beginning in 1953. That year the County of Santa Clara distinguished itself as a pioneer in the establishment of "exclusive agricultural zones" for the purpose of protecting some of the richest farmlands in the area.[109] Within a few years, however, the benefits of this project were lost, as these lands were polluted or developed during San Jose's annexation wars and the subsequent boom in the land-hungry electronics industry.

A third missed opportunity was the promise of pollution prevention, promoted by environmentalists and the USEPA and publicly endorsed by most large computer and electronics firms.[110] To be fair, the electronics industry has been successful at reducing many pollutants, if we measure those improvements in per unit production. However, progressive environmental initiatives in electronics have generally only occurred when grassroots activists or regulators demanded them, and as the volume of production has increased with the growth of markets, the total pollution burden has also increased, rendering any per unit pollution reductions moot.

Like El Pueblo de San Jose's unsustainable increases in crop production in the 1770s, the introduction of unnecessarily destructive mining technologies during the Gold Rush, and the continuous bouts with overproduction and water depletion in the canneries and orchards of Santa Clara County, Silicon Valley industries followed the time-honored practice of pushing past any reasonable limits on human and ecological sustainability, placing profits before people and the environment. In Santa Clara County there has always been a pervasive set of ideological and cultural tenets that view economic growth as unconditionally positive—an ideology foisted on the county's less influential majority by the most powerful classes. Just like the city planners' policy during the 1940–1970 period, Silicon Valley's prime directive today is "growth at any cost," despite the fundamental unsustainability of the approach.[111]

Whether from depletion or pollution, Santa Clara County industries seem always to have been intent on maintaining a dependence upon—and a lack of respect for—water, land, immigrants, and people of color. Environmental racism in the Valley meant not only that people of color were being exposed to toxics and pollutants at home and work, but also that this process was part and parcel of a broader context of general ecological degradation in the region. European contact, the missions, mining, farming and canning, and computer/electronics production each brought the promise of economic prosperity and new social liberties springing forth from the bountiful wealth of natural resources that only California could offer. But in each case, economic gains were concentrated among a few while poverty and immiseration were shared among the many; racial and ethnic cleavages reemerged and deepened; and the integrity of the natural environment suffered as yet untold assaults. Nothing is new about the latest proclamation of salvation in Silicon Valley; it is old wine in new bottles and represents only the most recent manifestation of a long history of environmental injustice, California style.

5

The Political Economy of Work and Health in Silicon Valley

> Nitric acid, one of the most heavily used chemicals in the electronics industry, is very nasty stuff. Inhaling even small amounts can kill a person, filling the lungs with suffocating fluid. If nitric acid gets in your eyes, they sizzle and shrink. Where it touches your flesh, the skin dies and eventually becomes black and shrunken, a process called "coagulation necrosis." If it splashes down your throat, you might literally vomit your guts out. That's not the kind of image most people get when they think about the high-tech industry.
>
> —Christopher Cook and Clay Thompson, "Silicon Hell"

Introduction

While several writers and scholars have examined Silicon Valley's high-tech industries in recent years,[1] we present a much broader range of occupations in which low-wage and/or "unskilled" workers are concentrated there.[2] This includes the *core* jobs in semiconductor production and the *periphery* jobs in printed circuit board, printer, and cable assembly, including home-based piecework and prison labor.

By placing this story in an environmental justice framework, we contribute to the burgeoning effort by some scholars to center the workplace in EJ struggles.[3] This is important for several reasons. Historians have demonstrated that our knowledge of industrial toxins—the same knowledge that fueled the modern-day environmental movement—is rooted in medical studies of workplace hazards generated in the early twentieth century.[4] This point is crucial because it underscores what some contemporary environmental justice and labor leaders have often argued: (1) that industrial workers were among the first victims of, and the first to resist, environmental pollution; and (2) if the pollution from smokestacks imposed upon communities is significant, then the impacts on the workforce inside the factories must be as great if not greater.

We also integrate into environmental justice studies a serious consideration of the role of gender. As with each of the previous economic systems that predominated in the Santa Clara Valley region, gender plays a major role in Silicon Valley's contemporary environmental justice struggles.

We draw our data from a wealth of archival, documentary, interview, and first-hand ethnographic sources, as well as legal depositions, to present a portrait of the Silicon Valley workplace like none other.[5] Building on previous chapters, we argue that contemporary environmental inequalities in Silicon Valley's electronics industry are the result of a combination of (1) historical patterns of labor migration and the concentration of a multi-ethnic, immigrant, and non-union workforce in the region; (2) industry's introduction of environmentally destructive technologies and authoritarian control over the labor process; and (3) the degree of autonomy, organization, and resistance among workers and social movement organizations. These are the underlying factors in the continuing battle over private profit, good jobs, and natural resources in Santa Clara County, which began in 1769.

Silicon Valley Industry: Its Size, Scope, and Workforce

While it has long been acknowledged that the United States has largely shifted from a goods-producing or manufacturing economy to a service economy, manufacturing remains key to the nation's fiscal health. The largest manufacturing industry in the United States and the world is the electronics sector, and "because of its growth and size, the chip industry is the pivotal driver of the world economy."[6] In 2000, U.S.-based electronics companies employed more than 280,000 people worldwide (up from 265,000 in 1998), and all electronics firms employed 2.4 million persons worldwide. Globally, electronics is a $300 billion industry. Intel, the world's largest chip maker, is also one of the world's most profitable companies, with a market value of $117.6 billion in 1998—more than the Big Three auto makers combined.[7] More than 220 billion microchips are manufactured annually.[8]

Electronics is a highly competitive industry that includes the production of semiconductors, microchips, disk drives, circuit boards, consumer electronics, communications devices, and video display equipment. The manufacture of high-tech electronics goods has also spurred an extraordinary increase in those industries that produce the materials and chemicals that supply electronics firms and those companies that treat, recycle, and dispose of hazardous waste generated in the electronics production process.[9]

By 1986, the computer and semiconductor industries made up 50 percent of Silicon Valley's manufacturing employment. And though many Silicon Valley firms are shifting production jobs offshore, to southeast Asia, Latin America, and Europe, research and development and specialized production have maintained the Valley's position as the primary high-tech center in the United States.

The electronics industry is often associated with outrageous fortunes and seemingly contagious wealth creation. In 1999, the CEO of Hewlett-Packard was paid $69.4 million, and Santa Clara County has one of the highest numbers of millionaires per capita of any community in the world. Unfortunately, this wealth is being siphoned upward; hundreds of thousands of production workers and residents are struggling to stay out of poverty. In 1998, "after taxes, a low-skilled worker earning minimum wage brings home less than $10,000 a year in Silicon Valley."[10] In an effort to ease tensions between labor and management, the San Jose City Council adopted the nation's highest minimum wage in November 1998. City contractors were required to pay at least $9.50 per hour with health benefits and $10.75 if benefits were not provided.[11]

Low-wage workers (those in the tenth and twenty-fifth percentiles) have seen their earnings plummet by more than 13 percent since 1989. The increasing discrepancy between "executive compensation" and the wages of lower-tier workers further exacerbates these inequalities. Between 1991 and 1996, for example, the ratio of annual income of the top one hundred Silicon Valley executives to the average production worker jumped from 42:1 to 220:1. About 20 percent of the Valley's jobs do not provide a living wage (defined as the level of earnings that would enable a single individual to achieve self-sufficiency). And almost 55 percent of the area's jobs do not offer enough pay to maintain a family of four above the poverty line.[12]

Income inequality is compounded by the exorbitant cost of living in Silicon Valley, which is now legendary. In 2001, the average cost of a home in the county was around $450,000. As one Chicano minister-activist in San Jose told us, the skyrocketing costs disproportionately impact people of color and immigrants:

> Some people are being forced to share a house or apartment with other people. For instance, in the church where I used to work, there are two families to a house, a three-bedroom house occupied by 16–18 people: two bathrooms and a small kitchen. [For that] we're looking at $1,600–$1,800 a month. They make ends meet in that way, not having too much space but paying something somewhat affordable. The people in my church are 90 percent Latinos, not at poverty level, but not middle-class level, somewhere in between: unskilled laborers. Enough of them are doing two or three jobs, working sixteen, eighteen hours a day so they can make the payments on the rent and buy the rest of the things they need. [By comparison to this area] in Fresno I was paying $750 for a new three-bedroom house.[13]

Inequalities in the Workplace

An estimated 70 to 80 percent of people working in production jobs in the Valley are immigrants, women, and people of color. This is an important point because it gets to the heart of environmental inequality. We must establish the demographic profile of the workforce and also ask: how did they come to be concentrated in hazardous high-tech work?

Most of these workers emigrated to the United States seeking an improvement in their economic situation, and most lack the skills and social networks needed to move up and out of these jobs. San Francisco State University sociologist Karen Hossfeld has researched immigrant working women in Silicon Valley for more than twenty years. She once noted that an employer she interviewed revealed his formula for hiring production workers: "He told me, 'There's just three things I look for in entry-level hiring. Small, foreign and female. You just do that right and everything else takes care of itself'. Employers assume that immigrants will work for less and that women are second wage earners."[14] Much of the ethnic and gender segregation in the industry is the result of conscious and selective recruiting on the part of managers.[15] "High technology personnel managers routinely link the broad societal images of 'passive Asians,' 'desperate Latinos,' and 'militant blacks' to help them hire individual workers and create a work force that they believe is less likely to demand rights individually or organize collectively."[16] An age-old myth states that women workers are supposed to have "nimble fingers" and are more skilled than men at performing the intricate work required in electronics firms. As one manager put it, "they're just so damned good at it!"[17] According to a spokesman for the National Semiconductor Corporation, "Experience has shown that women seem to be more dexterous and generally better suited for this kind of work. The young men tend to get antsy."[18] This is especially believed to be the case with southeast Asian women. As some industry insiders used to put it, the "secret weapon" in the "competitive field of sophisticated electronics devices is the 'FFM'—or 'fast fingered Malaysian.'"[19]

Selective recruitment often works against certain groups. As one activist told us, "African American women are angry at the temp agencies because the receptionists are polite on the phone when they schedule an interview, but when they see them in person [and realize they are African American], they tell them that there are no jobs available. This is invisible discrimination."[20] One of the authors had a similar experience when applying for an assembly job at a chip-making firm in the Valley. All was fine over the phone when scheduling an interview. But once the author arrived and filled out an application, the supervisor appeared from around the corner, looked him over from a distance, and turned away. Her administrative assistant emerged from the office just seconds afterward and told the author "my supervisor is busy in a meeting right now and won't be able to see you today." Given his phenotypic appearance as an African American male, the author could discern no reason for this behavior other than the employer's selective eye for "small, foreign, and female" workers. In short, Silicon Valley businesses deliberately select a largely "immigrant female of color" workforce to perform the task of production in the most toxic occupations.

As in previous eras, people of color, particularly immigrant women, continue to form the backbone of Santa Clara County's economy. During the 1970s, Asians held a modest segment of all semiconductor jobs (5.6 percent) and approximately 25 percent of all minority employment.[21] During the 1980s, however, Asians increased their share of those jobs to nearly 16 percent and their share of overall minority employment to nearly 50 percent. In less than a decade, Asian immigrants replaced Latinos as the single largest group of people of color in the industry.[22] Latino women were the largest single demographic in the industry in 1980, but by 1990 Asian immigrant women had become the statistical majority. Within and beyond electronics, Asians and Latinos are disproportionately concentrated in low-wage and/or high-hazard occupations in Santa Clara County. In 1992, Asians, who represented 17.5 percent of the county's population, held 57 percent of electronic equipment assembly jobs, and 26 percent of hairdressing and cosmetology jobs; they accounted for 24 percent of Santa Clara County's parking lot attendants and 21 percent of the cooks. Latinos, representing 21 percent of the county's population, accounted for 76 percent of farm worker supervisors, 45 percent of brick and stone masons, 53 percent of auto body repairers, and 57 percent of gardeners and groundskeepers in the county.[23]

As we noted earlier, the 1965 Immigration Act was responsible for much of the new Asian and Latin American influx into the United States. But more recently, the electronics industry has played a strong role in lobbying for further "liberalization" of these policies. The Immigration Act of 1990 was viewed as one of the most sweeping changes in immigration law in forty years. It opened up the doors for 140,000 workers to enter the country each year (up from 54,000 annually), the vast majority of whom were to be categorized as "professionals," "skilled workers," or "investors." This change was in large part initiated by the high-tech industry lobby, which had complained of a massive shortage of skilled workers.[24] What went largely unreported about this legislation was that many of these "professionals" and "skilled workers" could often only find jobs at low wages in the high-tech industry.

The massive influx of immigrants and people of color into the Valley since 1965 afforded the electronics industry a prime opportunity to exploit cheap labor. Romi Manan, a labor activist and electronics worker in Santa Clara County, had a great deal of experience with exploitation and racism in this sector:

> I have worked at National Semiconductor Corporation since 1979. I was hired as an operator on the swing shift. Discrimination at National Semiconductor is something I have experienced personally, something I have observed happen to others, and something which anyone can observe who wants to as a general phenomenon. The department in which I work was about 90 percent Filipino workers. Despite the overwhelming minority composition of this department, since I have worked there we have never had anything but white supervisors, white general foremen, and white production managers. That is the story generally at National, where production workers are largely Filipino, black, Hispanic, and other minority workers, and management is largely white, as are engineers and office workers. The supervisor seems to share the notion that Filipino workers will never protest or make waves. Many white supervi-

> sors at National think this way and take advantage of Filipinos who have difficulty with the language and don't know their rights under the law. A large portion of the Filipino workers at National are on the swing shift, and when cutbacks come this is where the company cuts first.[25] It's bitter to hear it, but we're just cheap labor for them.[26]

Michael Eisenscher, a labor activist in the Bay Area who worked with Romi Manan to organize National Semiconductor workers during the 1980s, also critiqued the racial division of labor and exploitation in the electronics industry:

> Depending on the company you went to, you would see whole departments composed almost entirely of Filipinos, another department primarily Vietnamese, a night shift would be more people of color. Very careful patterns, not just between job categories but within them. Then there were patterns in which company A would hire almost entirely southeast Asians. Company B would hire mainly Latinos. You had segregation between corporations and facilities, by occupation, by shift, by job assignment, right across the board.[27]

However, many managers at high-tech firms will insist that Silicon Valley still represents a land of opportunity: "There is a steady stream of people seeking jobs in entry-level manufacturing, and they are coming from the fast food industry or retail," Gary Burke of the Santa Clara Manufacturer's Association claims. "They are glad to increase their wages by about a third. The fact is that those entry-level manufacturing jobs have provided a job and a career ladder for people as they improve their skills."[28] The data we present here offers a significant challenge to these claims.

Organizing against Toxics: From the Workplace into the Community

Today, up to a thousand different chemicals and metals are used in the various processes required to produce semiconductor chips. Clearly, the toxic spills that have enraged communities located near chip plants must also be impacting those workers inside the plants. In chapter 4, we discussed the now infamous discovery of solvents (TCA) leaking from the

Fairchild Semiconductor Corporation into nearby soil and water and into the bodies of Los Paseos neighborhood residents and their children. Inside a typical semiconductor plant, workers are exposed to levels of TCA that, while legally permissible, are "substantially higher" than the exposure experienced by the residents of San Jose whose drinking water was tainted with the chemical. Moreover, many Fairchild workers received a "double dose" of TCA, because after being exposed to the solvent on the job, they came home and drank water that was also contaminated with it.[29] The joint contamination of the workplace and the community was finally clear. Underscoring the importance of workplace-community environmental links, Frank Mycroft, director of the Toxicology Information Center at San Francisco General Hospital's Poison Control Center, stated that "any attempt to stop environmental exposure to toxic chemicals must start in the work place."[30]

Social Movement Organizations Emerge

Earlier, we introduced two of the main environmental justice organizations in the Valley—the Santa Clara Center for Occupational Safety and Health (SCCOSH) and the Silicon Valley Toxics Coalition (SVTC). Here we consider many of the struggles these groups initiated, won, and continue to fight in Santa Clara County and beyond. These organizations are the most visible forces in the effort to make clear the links between toxic workplaces and contaminated communities.

SCCOSH cofounder Amanda Hawes is a workers compensation lawyer with a background in legal aid. She began her work in the region in the early 1970s, representing workers seeking better conditions in the canneries.[31] SCCOSH was established in 1978; at that time it was a combination of two projects: Electronics Committee on Safety and Health (ECOSH) and the Project on Health and Safety in Electronics (PHASE). These groups led the effort to monitor Silicon Valley industries and achieved a number of accomplishments: they formed the Silicon Valley Toxics Coalition (SVTC) in 1982; started a support group called Injured Workers United; helped found the Occupational Clinic at Valley Medical Center; created a Hazards Hotline for Electronics Workers; and launched the Campaign to Ban TCE (trichloroethylene) as a reproductive hazard.[32] Hawes, frequently emphasizes the workplace-community environmental links:

> A big success for ECOSH was our . . . campaign to ban TCE. The TCE campaign was noteworthy because it brought the spotlight on the chemical-handling aspects of the so-called clean industry at the same time TCE contamination of local water supplies was coming to light. But it was mainly noteworthy because *its origin and focus was always the workplace. If hazards faced by workers are not made a priority, we will all suffer the consequences.*[33]

Hawes not only links the workplace and community with regard to environmental contamination, but she explicitly argues that these problems originate in the electronics plants. This is an important observation because it underscores the fact that workers are on the front line of toxic exposures in Silicon Valley.

SVTC's executive director, Ted Smith, echoed these sentiments while speaking to a group of Valley workers and environmentalists in June 2000. SVTC has been at the forefront of the movement to recognize the widespread toxic contamination in the Valley as a form of environmental injustice:

> SVTC was founded in the wake of the Fairchild chemical spill, which created a lot of cancers and birth defects. We organized around that case and that's how this movement really got started. We started out as a project of SCCOSH and we were a spin-off. We discovered 29 National Priorities List toxic sites and they are often in industrial areas where people of color, immigrants, and the poor live. This is environmental injustice and we have worked with SCCOSH to highlight these *problems in the communities and in the workplace because the exposures are the same.*[34]

While the workplace-community environmental links are evident, what is disturbing is that workers in hazardous jobs receive higher doses of chemical exposure than do residents living nearby. This occurs not only because workers are, by definition, in closer physical proximity to the hazards, but also because federal law allows it. As one report revealed: "Occupational exposure limits are the highest legally-permitted levels of human exposure to toxic substances. Environmental exposure of the public to the same air pollutants is regulated much more strictly. Workers' exposure to toxic substances should be no greater in the workplace than in other regulated settings of human exposure to toxic substances."[35]

The links between environmental pollution inside the plants and the impacts on the outside communities are crystal clear in Silicon Valley. For example, more than twenty of the Valley's thirty-one toxic Superfund sites are directly related to the electronics industry's pollution. In 1997, a study conducted by the California Birth Defects Monitoring Program found that women who lived within a quarter mile of a Superfund site during the first three months of pregnancy were four times more likely to give birth to a baby with serious heart defects, and two times more likely to give birth to a baby with neural tube defects (e.g., spina bifida and anencephaly) than women living in communities with no Superfund sites.[36]

Despite the highest density of toxic waste sites in the nation, with workers dying of cancer in the plants, and with babies in surrounding communities suffering birth defects, industry representatives have maintained that no relationship exists between the heavy chemical use in electronics and these elevated negative health outcomes. Workers and residents have worked to challenge this claim. As the great activist Mother Jones was famous for saying, "don't mourn, organize!" When an entity as powerful as the $300 billion electronics industry is not "convinced" by well-researched reports and studies of illness and death in their back yards, the use of political power is perhaps the most effective way to persuade them to take notice.

Given the deep tradition of anti-unionism in the Valley, SCCOSH was one of a few worker-based organizations advocating the type of changes for which unions might normally struggle. Under Eula Bingham's directorship, the federal Occupational Safety and Health Administration's (OSHA) New Directions Grant Program (which began in 1978) gave financial support to COSH groups throughout the United States. SCCOSH's mission was to educate workers, health officials, and the public about the nature and extent of hazards associated with the electronics industry. OSHA supported SCCOSH's PHASE (Project on Health and Safety in Electronics) project because it focused on education and providing technical assistance to workers. As SCCOSH cofounder Robin Baker remembers, "Most grants went to major international unions, but Eula Bingham wanted to reach under-served workers, including women, people of color, and the unorganized. None of us had ever written a federal grant before, but we wrote a successful proposal that established and funded PHASE."[37]

The founders of SCCOSH were three women, each of whom was, in some way, connected to the electronics industry. Amanda Hawes was an attorney representing electronics workers in compensation cases, Robin Baker worked for a local university's occupational health program, and Pat Lamborn was an activist and electronics worker. As we noted earlier, it is often the case that women fill the leadership roles in grassroots environmental and public health organizations around the world.[38] Historians have documented that women have often strategically framed their community activism as "municipal housekeeping," an unthreatening extension of their traditional role as domestic caretakers.[39] However, SCCOSH and other area grassroots groups have more boldly defined their mission as education and agitation, and often in feminist terms.[40]

SCCOSH faced many constraints and challenges from the beginning. Federal funds could not be used for organizing workers, so SCCOSH's founders restricted their political activities to the PHASE project. PHASE quickly became a target of industry, which suspected that it operated as a "front" for union organizing. In 1980, safety engineers from twenty-five companies lobbied OSHA officials to end the agency's funding of PHASE. These engineers also believed, because of their faith in the industry's alleged "clean and safe" reputation, that an organization like SCCOSH served no useful purpose.[41] In November 1981, the office of the U.S. Secretary of Labor sent a letter to SCCOSH announcing that the funding for PHASE would not be renewed and that the group had until December of that year (one month's time) to finish its work. This decision came despite the obvious need for PHASE's Hotline and other services.[42] Robin Baker remembered:

> When Reagan was elected president in 1980 and Thorne Auchter became the head of OSHA, the Labor Department launched a crackdown on all labor-friendly activities. PHASE was the first to go. OSHA informed us that semiconductor industry reps had hired private detectives and prepared dossiers on Pat, Mandy and me, as well as PHASE activities.[43]

SCCOSH was in a long line of movement/community organizations that threaten powerful institutions and therefore were targeted for surveillance and dismantling.[44] In response to this withdrawal of funding,

> We decided to start a sister group, ECOSH [Electronics Committee on Safety and Health], to do what we could not do with federal funding—that is, help workers advocate for better health and safety conditions. SCCOSH was formed in 1978 as the umbrella organization for both ECOSH and PHASE. We chose the name Santa Clara COSH to link up with the COSH movement nationwide. The support of this national movement was invaluable.[45]

In this way, SCCOSH began to move toward a stronger political stance after industry and the federal government attempted to undermine the organization. SCCOSH became bolder and began to collaborate more aggressively with workers in the plants and with sympathetic health researchers:

> A researcher from the University of California who needed help locating women electronics workers exposed to TCE (trichloroethylene) came to ECOSH. The study offered the potential for health protection in the prevention of breast cancer, but when ECOSH members put flyers in the women's restrooms at several semiconductor plants, management immediately removed them. Then a flyer was sent to the *San Jose Mercury* (newspaper) and the search became news. Within 24 hours ECOSH's broom-closet-sized office in Mountain View was deluged with calls from hundreds of women electronics workers anxious to participate—recounting stories of their own exposure to TCE and other chemicals, and health problems in themselves, coworkers and family members. This outpouring of concern over health in the clean rooms convinced ECOSH of the need for a center that could gather and disseminate health hazard information to electronics workers and could advocate for improved working conditions in the so-called clean industry.[46]

SCCOSH and SVTC joined with a number of other organizations and individuals working on different fronts of the struggle to clean up the Valley and make it safe for workers and residents. For example, Dr. Jim Cone, a medical pioneer in the effort to bring attention to the occupational health issues in the Valley, stated:

> I started with this work twenty years ago in Redwood City where I saw a lot of workers who were dying from lung cancer. Many scientists like myself have since raised a lot of questions about the electronics industry

> and its workers because of the lack of protective standards for employees. Much of the problem is the opposition from the chemical companies. Many volunteer scientists got together with SCCOSH to produce books, pamphlets, and Internet information that have been read by workers all over the world. We are saying that it is not OK to poison workers. So far, workers have not been allowed to participate in these discussions. Instead, scientists and industry have had conversations about what the maximum allowable exposure levels can be, when we should be talking about what safe levels are.[47]

From the beginning Cone has been an ally in the movement to expose the secrets of Silicon Valley. He is also a respected occupational health scientist, and his credentials brought added credibility to the claims activists were making about the electronics industry.

Asian Immigrant Women's Advocates (AIWA), another activist organization, has been working in the region since 1983. AIWA is a community-based group that organizes immigrant women working in low-wage industries such as the garment, electronics, hotel, restaurant, nursing home, and janitorial sectors in the Bay Area. As the name implies, this organization is focused on the reality that a significant percentage of production workers in the area are Asian immigrant women. Raj Jayadev, a Silicon Valley activist and author who has worked with AIWA, had this to say:

> What they do is, they will go to a plant and talk to workers. For some reason, people don't do that these days. That's how easy it is. All you have to do is go and wait in front of a plant when there are shift breaks and try to talk to folks as they're leaving. That's how they keep it real. They know exactly what's going on, which companies move, who's working where. The organizers will leave their office space and do that. Now they've done it so well that they have women workers in each one of them, like IBM or whatever, where those women can do outreach.[48]

AIWA, SVTC, and SCCOSH have pursued activities traditionally dominated by unions. Advocacy for immigrants, particularly the undocumented population, has been a large part of SCCOSH's mission as well. Given the prevalence of community-based organizations involved in political activity directed at the electronics industry, one wonders where the unions are. The relationship between the environmentalist-occupational health or EJ alliance and labor unions in Silicon Valley has almost always been tenuous at best. Unions are essentially nonexistent in the electronics sector in Santa Clara County, the result of a long history of union-busting by the industry.[49] Those few unions that have tried to enter the Valley have taken a step that most labor organizations would shy away from, because of the power of Silicon Valley industry and because of the wide gulf between the traditionally European American male–dominated trade unions and a Valley labor force that is mostly female, immigrant, and people of color. Distinguishing itself from the pack, the United Electrical Workers Union (UE) made continuous efforts during the 1980s and 1990s to organize Valley workers. They were actually successful at organizing the first electronics-based union ever in the Valley, in 1993 at Versatronix. Michael Eisenscher is a long time UE activist who was sent into Silicon Valley in 1980 to organize workers:

> The cost of living in the Valley was going crazy during a period of national inflation. There was tremendous pressure on workers' paychecks, which were low to begin with. We began a campaign around cost of living increases. While they wouldn't acknowledge it, we extracted from the industry the first general wage increase, and once we broke the back of the issue at National Semiconductor, the others followed suit. We also conducted a campaign on the issue of discrimination, demanding that what we could see occurring in the plants—the patterns of segregation and tracking people by race and ethnicity—be brought to an end. So people were in a position of powerless subservience. We have been arguing all along that this "high-tech, modern industry" is nothing but warmed-over capitalism. It's the same. There's nothing remarkable about it.[50]

With all of the movement organizing occurring in the Valley, cracks have appeared on the industry's public image from time to time, and company policies and state laws have changed as a result. One of the first steps the environmental justice movement in the Valley took was to prove that there was indeed a problem with the electronics workplace.[51] This involved encouraging government, industry, and health officials to conduct health studies of the workplace.

Worker Health Studies

Occupational injuries, illness, and fatalities continue to have major impacts on the state of public health in the United States. In 1997, more than 6,200 persons suffered workplace-related fatalities; between six and twelve million workers are injured or experience a work-related illness every year. Moreover, people of color, immigrants, and the working class suffer a disproportionately high percentage of these workplace environmental hazards—what many observers have called environmental injustice. The concentration of people of color in hazardous, health-compromising jobs is well documented.[52] However, the relegation of women, particularly immigrant women and women of color, to hazardous jobs is a core component of this process and deserves equal attention. First, however, we consider the general association between social inequality and high technology.

Journalists and politicians have paid much attention to the issue of the "digital divide"—the social gaps in access to high technology across race, class, and gender. This gap has not only stayed with us, it has widened—another broken promise of the high-tech revolution. For example, in 1991, white children in the United States were 2.5 times more likely to have a home computer than were African American and Latino children.[53] With regard to gender, the promise from industry leaders early on was that occupational segregation in high-tech jobs would become a thing of the past, given the need for "brains not brawn" in the new economy. No such gender equality has occurred. According to one study, "there is more occupational segregation in industries classified as high-tech than in those that do not rely so heavily on technology." Furthermore, "the higher the technological intensity of the industry, the greater the degree of occupational segregation by sex."[54] Those women who do make it to the top of the corporate ladder are few and far between. In 1992, Silicon Valley lagged behind the rest of the nation, with only 10 percent of the region's one hundred largest firms having any women at all on their boards of directors. This 10 percent figure is abysmal when compared to the 50 percent of the nation's one thousand largest companies that have at least one woman on their board. As with a range of social, economic, and environmental indicators, Silicon Valley also stands out as especially backward with regard to gender equality.[55] It is clear that at every occupational level in high tech, women face ongoing discrimination and a general lack of power. These dynamics are especially problematic for women in this industry, because they also face the highest risk of toxic exposure.

If a researcher were to draw on the U.S. Bureau of Labor Statistics data to evaluate the health risks in the electronics industry, s/he might find no cause for alarm. In fact, during the 1980s, those statistics indicated that high-tech workers had an accident and illness rate about half the national average. The main reasons for this lack of accuracy include the industry's deliberate manipulation of the definition of "illness" to reduce these numbers and the fact that many of these illnesses (cancer, for example) may not be recognized as work-related at all and often do not present themselves until decades after workers are exposed (and have left the job).[56]

Despite these data gaps, various researchers have conducted studies of this industry in Silicon Valley, along Boston's Route 128 high-tech corridor, in Scandinavia, and in Britain. Many of these health studies were initiated largely as a result of worker and social movement agitation around environmental and occupational safety issues in the communities surrounding electronics plants.

According to the State of California's "Occupational Injuries and Illness Survey" for 1977 and 1978, the rate of illness in the semiconductor industry was 1.3 per 100 full-time workers, while the rate was 0.6 and 0.5 during the same period for workers in general manufacturing.[57] In 1980, the difference increased to where there were three times as many illnesses among semiconductor workers as there were among the general manufacturing labor force. In 1981, the California Division of Occupational Safety and Health produced a study of forty-two California semiconductor plants. This report concluded that much (47 percent) of the elevated illness rates in the electronics industry was the result of systemic poisoning, or toxic exposure.[58]

Two later studies, sponsored by the National Institute of Occupational Safety and Health (NIOSH), also explored health-related issues in the industry. One of these studies, conducted by the Battelle Institute, found that the etching equipment commonly used in electronics production can produce exposure to radiation above legal standards. The authors of that study discovered that production workers can also be exposed to significant doses of arsenic and toxic gases under "normal working conditions."[59] A second NIOSH-sponsored study, by the Research Triangle Institute, underscored that the synergistic effects of chemicals used in the industry may place workers at greater risk than would be anticipated when using traditional health studies (based on risk assessment models) that only examine individual chemical exposure.[60] In other words, scientists were beginning to acknowledge that the tools they were commonly using to determine occupational health risks were not based on actual exposures. The reality in electronics plants is that workers are exposed to multiple chemicals at a time, something that traditional risk assessment models are unable to accommodate. This also meant that any health study of the industry that did not consider multiple chemical exposure might grossly underestimate toxic exposures at best, and be totally unreliable at worst.

The lethal nature of electronics production first became known in 1983, when a study published in the *British Journal of Industrial Medicine* found that electronics workers in Sweden, who had been tracked for ten years, had a higher than average incidence of all cancers, particularly those affecting the larynx and the respiratory system.[61]

The semiconductor industry quickly responded to this growing body of damning evidence with its own Occupational Health System (OHS), a different and supposedly more accurate way of classifying injuries and illnesses.[62] The Bureau of Labor Statistics had defined an injury as an event "which results from a work accident or from an exposure involving a single incident . . . occurring in one instant."[63] As there was no further clarification as to what constituted "one instant," the Semiconductor Industry Association (SIA) capitalized on this ambiguity and began recording "one-time" chemical exposures as "injuries" rather than illnesses. This produced an immediate and dramatic decline in the industry's illness rates.[64] The evidence of this drop was publicized in a report by a team of researchers at the University of California-Davis in 1987. They found that only 60 percent of all work-related injuries and illnesses suffered at semiconductor plants were being reported to federal occupational health officials. The findings underscored that the industry "does not completely or consistently record" illnesses with subtle symptoms or those that remain dormant for an extended period of time.[65] State officials soon thereafter forced the semiconductor industry to change its reporting methods to include a range of exposures that had been excluded from illness records. This deliberate manipulation of health data by the industry only reaffirmed the general pattern of companies taking steps to manage appearances and to create an image of peaceful coexistence with labor and communities.[66] This is at the heart of how and why Silicon Valley worked so hard to create a "clean" image of itself—to ensure its own survival above all else.[67]

The Movement Mobilizes around Health Studies

But Bay Area EJ organizations were making sure that industry would have to answer for its toxic workplaces, especially regarding solvents. As early as January 1981, SCCOSH published a technical report that cited a National Institute of Occupational Safety and Health (NIOSH) study documenting that the industry's commonly used solvents, known as glycol ethers, were associated with reproductive disorders in animals. A year later, the California Department of Health Services issued a health warning to California workers regarding glycol ether solvents, concluding that "glycol ethers have damaged the reproductive systems of test animals, raising the possibility that they may cause similar effects in humans." That same month (May 1982), the SIA issued an alert to executives in the industry regarding glycol ether health effects. A year later, in September 1983, the Chemical Manufacturers Association issued a Research Status Report on glycol ethers, documenting the extensive nature of their associated reproductive toxicity in animal studies. And in 1985, the California Department of Health Services released its epidemiological study on residents exposed to semiconductor solvents that had contaminated local drinking water as a result of the chemical spill by the Fairchild Corporation in San Jose. The study found that birth defects and miscarriages were occurring at 2.5 to 3 times the expected rates.[68]

Despite the workplace origins of toxics-related problems in the electronics industry, the U.S. Environmental Protection Agency—*not* the Occupational Safety and Health Administration—became more involved with regulating the industry.[69] In 1986, the USEPA issued a report urging that all 52,000 semiconductor industry employees in Santa Clara County undergo a comprehensive series of health tests to assess the effects of handling toxic substances. The report also revealed that 44.4 percent of the county's workforce held jobs in the area's 68 most toxic industries. The EPA charged that Cal-OSHA was too nonchalant about the need for eliminating toxic threats to workers. They also lamented the shortage of occupational health specialists in the county that year.

Specifically, the county health department had only one employee—a chemist—in its occupational health division, down from eight employees during the 1970s.[70]

Some industry leaders were concerned with these developments. Interestingly, however, the first computer corporation to sponsor a health study was in the Boston area, not Silicon Valley. In 1986, Digital Equipment Corporation (DEC) conducted a study that found that its female workers experienced miscarriages at nearly twice the national average.[71] All of these women worked with a variety of toxic liquids and gases used to manufacture semiconductors.

Prompted by the DEC study, in January 1987, AT&T, one of the world's largest semiconductor manufacturers, barred pregnant women workers from production lines in many of its computer chip plants in the United States—using what is commonly referred to as a Fetal Protection Policy.[72] Many labor and environmental activists responded to AT&T's announcement with disgust because it did nothing to address the fundamental problem of toxics in the workplace. "Ban the toxics, not the workers!" Ted Smith of SVTC had said at the time. Both DEC and AT&T offered work transfers for pregnant women and free pregnancy tests for others.[73] Movement organizations were mobilizing nationally to urge the industry to reform around this and related issues.

In February 1987, several national organizations involved in workplace health and safety, women's legal rights, and related struggles wrote a letter to the presidents of SIA member companies, protesting the use of fetal protection policies in place of pollution reduction in the workplace. Later that year, many labor, medical, industry, and community organizations participated in the "Semiconductor Health Hazards News Conference" in Florida. This conference was a catalyst for the SIA-sponsored study by the University of California-Davis that began in 1988.

In December 1992, the results of a three-year University of California study sponsored by the SIA were released. The principal finding was that pregnant women workers were still exposed to health risks in this industry, largely due to the use of the solvent glycol ether. Specifically, the several thousand female workers at 14 different chip plants throughout the United States showed a 20 to 40 percent greater chance of miscarriage than employees in the most toxic jobs.[74] Another finding—rarely reported—was that *both men and women* in the study were documented as experiencing higher than expected rates of respiratory problems and eye and skin irritation.[75]

Activists were angered when, in the wake of the study, SIA industry members failed to commit themselves to any particular timetable for the elimination of glycol ethers. SVTC Executive Director Ted Smith charged, "the semiconductor industry has shown that it cares more about the next generation of computer chips than the next generation of children."[76] Nancy Lessin, director of the Massachusetts Committee on Occupational Safety and Health (MassCOSH), argued that transferring women out of glycol ether–intensive production failed to solve the problem *for male workers*, who might also suffer from reproductive and other illnesses. "If they choose to remove workers, they should look to men too."[77] Silicon Valley EJ groups went so far as to threaten a boycott of all chip makers who failed to phase out glycol ether within a year's time.[78]

In October 1992, the preliminary results of a five-year study at IBM (conducted by Johns Hopkins University) were also released.[79] In that study, 10 of 30 pregnant women exposed to glycol ether at an IBM plant experienced miscarriages. The SIA recommended to its member firms that they accelerate efforts to eliminate the use of ethylene-based glycol ethers (EGEs).[80] While environmental justice activists were pleased at this turn of events, they were also disappointed because many of them had recognized this problem and had called for a phase-out of EGEs long ago. Amanda Hawes, executive director of SCCOSH, charged that the industry "has been stalling the adoption of safer alternatives until there is a body count."[81]

These studies shook the semiconductor sector and challenged the three-decade-old myth that this was a "clean industry."[82] Activists viewed these scientific affirmations of their claims as both a vindication and a window of opportunity for organizing to reform the electronics industry.[83] For example, in response to the 1992 SIA study of miscarriages, SCCOSH and SVTC announced the launching of the Campaign to End the Miscarriage of Justice, calling on the industry to phase out glycol ethers and other reproductive toxins. Representatives from this campaign led workshops on environmental health and safety at the annual conferences of the American Public Health Association and the Committees on Occupational Safety and Health (COSH) in San Francisco in 1993. In 1994, after campaign members met with SEMATECH, the taxpayer-subsidized consortium of chip makers, the industry began phasing out its use of ethylene-based glycol ethers.[84] Thus, social movement organizations had a major influence on regulators, universities, and the industry in that their protests and negotiations led to the commissioning of several health studies and the nationwide change in company policies regarding the use of certain toxins.

The struggle continues. Despite cosmetic reform, the industry remains fundamentally hazardous. In addition, because many chemical-induced illnesses take years to develop, persons who worked at a plant during the 1970s and 1980s may not present with an illness until 2000 or later. And when that happens, companies will fight to protect themselves from liability. This was the case when attorneys filed a wrongful death lawsuit against IBM in February 1998 in Santa Clara County Superior Court, citing a high incidence of cancer-caused fatalities among IBM employees. The suit was brought on behalf of the families of five former IBM workers who died of cancer, as well as four other current or former IBM employees suffering from cancer. In an innovative legal move, the suit also included the names of the chemical companies who manufactured the toxics IBM used. These include Shell Oil Company, Union Carbide Corporation, Shipley, Hoescht, and a number of other chemical manufacturers. The suit stated, "Motivated by a desire for unwarranted economic gain and profit, defendants willfully and recklessly ignored knowledge . . . of the health hazards. . . . The objective of these defendants was maximizing production, but in doing so, these defendants endangered the health and safety of IBM workers."[85] The *San Jose Mercury News* also reported in 1997 that a longtime IBM chemist, Gary Adams, who himself has fought cancer, alerted IBM management to the problem of carcinogenic chemicals in the firm's production process as early as the mid-1980s but was ignored.[86]

The Toxic Treadmill

Why, then, is the electronics industry so toxic? And why have industry leaders been so reluctant to restructure production so as to reduce toxic exposure? As we've discussed, the roots of this industry's chemical base are in the military, petrochemical, and agricultural sectors—the biggest polluters on earth. Furthermore, many of the contracts between military clients and electronics firms have traditionally required the use of "state of the art" technology, which often means highly toxic chemicals. Perhaps a more intractable problem is the constant drive to accelerate production and consumption—a drive that is fundamental to capitalism itself.[87] The manufactured desire for more rapid economic returns and faster, more powerful consumer products has perhaps presented the most daunting barrier to environmental reform efforts in this industry. Dr. Myron Harrison, a former physician at IBM, recently wrote of the inherent conflict between the imperative to continuously accelerate the computer production cycle and the need to protect the health of workers:

> Professionals associated with [semiconductor manufacturing] have invariably commented on the rapid pace of change in tools and materials and on the fact that adequate toxicological assessment of chemicals almost never precedes their introduction into manufacturing settings. The pace of change is quickening under the pressure of severe economic competition. As recently as 3-4 years ago, a typical schedule of a new technology from research and development to pilot lines to full manufacturing was 6-8 years. Executives who manage micro-electronics businesses are now demanding the schedule be compressed into a 2-3 year time frame. Engineers are not evaluated nor rewarded on their ability to . . . understand new or unusual health hazards. This task is the responsibility of health and safety professionals. Unfortunately, the opportunities for the professionals to be involved before these new processes arrive at the manufacturing floor are being diminished by the quickening pace of technologic change. . . . Any large semiconductor facility uses several thousand chemicals. Any attempt to review the toxicology of all these materials is doomed to be superficial and of little value. . . .[88]

A clear illustration of Dr. Harrison's observation is "Moore's Law"—a prediction (made by Intel founder Gordon Moore in 1968) that computer chips will double their speed and power every eighteen to twenty-four months. Since that time, Gordon Moore's status has grown to the level of icon, in large part because his prediction came true. Unfortunately, it is often the case that the greater the power and speed of electronics devices, the greater their toxicity. This relationship is borne out in the example of arsenic and gallium in the industry. These two chemicals are being used in "ever larger quantities because higher speed micro-electronic devices often require wafers composed of equal parts arsenic and gallium."[89]

More generally, the relationship between the speed and size of microchips and toxics is disturbing:

> Building smaller and faster circuits . . . requires the use of more solvents and other chemicals to achieve the necessary requirements for "clean" components. As the geometries of production decrease [as the size of the chips gets smaller], more solvents are needed to wash away ever smaller "killer particles" that could jam a circuit. Smaller and faster may also mean using even more toxic chemicals.[90]

Industry leaders generally claim that the chemicals in the industry are so new that no one can possibly know what the health outcomes of working with such cocktails of compounds might be.[91] In response to that claim, Richard Youngstrum, health and safety coordinator for the International United Electrical Workers Union during the 1980s, told a reporter, "You talk about the high-tech industry being 'new' but in fact many of the things happening on the industry are not new at all. . . . They're tried-and-true industrial processes that we know a great deal about already."[92] Other medical experts on the industry, such as Dr. Thomas Kurt of the North Central Texas Poison Center, agree. He argues that few *new* chemicals are being introduced into high-tech production. Frequently, these so-called "new" chemicals simply have not previously been used extensively, or are combinations of two chemicals, such as gallium-arsenide, whose individual effects are well known. Kurt says it is not accurate to claim that what these chemicals are going to do is totally unknown. "Almost all of them have been studied in other applications before the semiconductor industry emerged."[93]

Of the many hundreds of chemicals and compounds used in the electronics industry, many do support Kurt and Youngstrum's contention. For example, cyanide, arsenic, nitric acid, toluene, hydrochloric acid, xylene, silica, TCE, lead, and asbestos are all known human and/or environmental health hazards. Cyanide, an infamous poison, has been used in gold processing in California since 1890, and lead, asbestos, and silica have long been known to cause occupational diseases whose lethal effects on human health have been documented since Biblical times.[94] And in 1972, years before the first publicity emerged about the electronics industry's toxic workplaces, the International Labor Office in Geneva, Switzerland, reported that the chemical xylene irritated both blood-producing organs and the central nervous system.[95] Even so, many activists argue that in fact most chemicals used in high tech have not been adequately studied, particularly for their health and environmental effects.

TABLE 5.1

*Percent Work Loss Due to General Illness/Injury in Silicon Valley**

	1992	1993	1994	1995
All Mfg	2.1	5.8	6.0	6.0
Electronics	8.1	8.0	10.0	9.8
Semiconductors	8.5	9.2	11.0	12.8

* Bureau of Labor Statistics, United States Department of Labor, April 1997.

Table 5.1 reveals that the electronics industry in general, and semiconductor manufacturing in particular, are both associated with higher than average worker-days lost due to illness and injury. The bad news for U.S. workers is that, for all categories of manufacturing work, the amount of work lost to injuries and illnesses has increased. The news for electronics and semiconductor workers is worse, because their rates of work loss are much higher, with semiconductor workers' rates more than double the rate for general manufacturing employees.

Industry's Response

Aside from the modest effort to phase out glycol ethers, the responses from industry to the mounting evidence that their plants are toxic have included issuing categorical denials, providing misinformation, withholding information, harassing and retaliating against workers, maintaining a veil of secrecy, and refusing to cooperate with government authorities.

The first line of defense for the industry has always been the claim that the electronics sector is clean and safe. According to the Semiconductor Industry Association (SIA), the chip industry has ranked in the top 5 percent for workplace health and safety among U.S. companies since 1972. "These chemicals are tightly controlled. Our environmental health and safety program is second to none," asserts George Scalise, president of the SIA. "There is no possibility of exposure." A spokesperson for National Semiconductor, a firm targeted by a worker lawsuit concerning occupational illnesses, commented that the suit "is absolutely without merit."[96] "There is no scientific basis to justify a study," claims Lee Neal, health and safety director for the Semiconductor Industry Association. "We use chemicals in work environments two to three times cleaner than the typical operating room."[97]

Other officials were no less bold and sweeping in their efforts to engage in damage control. "This industry is the epitome of where you would want American industry to be," contends Donald Lassiter, a health consultant to the high-tech sector.[98] One manager of a plant who provided tours to a researcher studying semiconductor clean rooms claimed that many of his employees suffering from allergies and chronic asthma preferred to come to work because the air in the clean rooms was so pure.[99]

When the industry was facing its first onslaught of bad press regarding toxic spills and worker health problems, Thomas Hinkelman of the SIA charged that these reports were being drummed up by "people associated with attempts to unionize the industry." Furthermore, in his opinion, the industry "ranks among the best" in the world in health and safety.[100] In response to the IBM/Johns Hopkins study released in 1992, IBM spokesman Jim Ruderman stated, "The primary motivation for the study was to try to clear the tarnished reputation of semiconductor clean room health risks for women after the Digital [DEC] study. If there are any bright spots here, it's that the rest of the operations in our clean rooms are safe."[101] IBM spokesperson Tara Sexton responded to the lawsuit filed by employees suffering from cancer with the customary denial: "We do not believe that any of these illnesses was caused by anything associated with work at IBM."[102] Sexton also stated that "IBM has a longstanding commitment to a safe working environment, and compliance with all health and safety regulations and laws."[103]

In the event that ill or dying workers can prove the relationship between toxic work and their health, many firms have employees sign away their right to sue the firm. During the deposition of one plaintiff suing a major Silicon Valley firm for allegedly causing his cancer and other health problems, the company lawyer reminded him that he had signed a covenant not to sue: "Do you have a recollection of signing a general release or other document?" The attorney also reminded the plaintiff that the company was still paying his monthly pension.[104]

If public relations ploys and covenants not to sue fail to dissuade activists, ill workers, and regulators, companies often refuse to cooperate. After all of the studies of semiconductor workers that focused exclusively on miscarriages, environmental and labor advocates and the

USEPA tried to gain the semiconductor industry's approval to research the broader health impacts on employees with regard to diseases like cancer and birth defects. But, according to one report, "the high-tech industry has declined to cooperate."[105]

When workers and lawyers go after high-tech firms in court, withholding information, deceit, and outright lies are common tactics found in the corporate arsenal. For example, in a lawsuit that several women filed with Amanda Hawes against AMD (Advanced Micro Devices), the company repeatedly stalled when asked for documents, falsely claimed that certain data that workers regularly recorded on the job were never collected or maintained, and claimed that information they had in their possession was lost.[106] And when it was clear that certain workers were determined to pursue these lawsuits, retaliation was common. Shortly after one worker became a plaintiff in a class action suit against a large firm, her husband was laid off from the same company.[107]

Aside from the focus on women in the majority of these high-tech health studies, anyone studying gender politics in Silicon Valley cannot help but notice the gendered nature of corporate public relations. Many electronics firms that have come under fire for various social and environmental infractions use women as their company spokespersons. Lydia Whitefield was the spokesperson for AT&T Technology Systems in New Jersey in 1987, the year that company banned women from semiconductor chip production lines in response to the Digital Equipment Corporation miscarriage study; Marcy Holle was IBM's spokesperson in 1993 when they announced an end to their fifty-year traditional "no layoff pledge"; Tara Sexton was the spokesperson in 1998, when IBM was facing a class action lawsuit from former and current employees dying of cancer; Lisa Werne was Solectron's spokesperson in 1999 when the *San Jose Mercury News*'s investigation of home-based piecework prompted a public outcry against Solectron and other firms involved in these practices (see chapter 7); and Gretchen McWhorter and Jan Butler were IBM spokespersons when the company announced that it would be downsizing 1,500 jobs in 2001.[108] Each of these women was quoted in local and/or national media outlets during these crises. Women are often placed in the position of spokesperson for companies that are likely to face public criticism from time to time, because they allow the firm to present a "kinder, gentler" face, behind which mostly men are making the decisions and benefiting economically and politically. When a company has to deliver bad economic news or respond to charges of corporate malfeasance, the face and voice of a woman spokesperson is often believed to have a more soothing and politically calming effect on a potentially rebellious public than a man's image and voice. In this way, women have become central to the production of both high-tech capital and the ideological and rhetorical symbols that support that system, even as they have also borne the brunt of this assault on labor and the environment.

6

The Core

Work and the Struggle to Make a Living without Dying

> You discover that all those nice things they told you about how they were protecting you were just bullshit. They got rid of me because I was vocal. I couldn't be intimidated. That bothered them. You have to speak up. No job is worth my health. I cared more about my safety than about the wafer [the microchip]. And that's a basic conflict.
>
> —Nancy Hawkes, former Advanced Micro Devices (AMD) employee and cancer survivor

Core Occupations in Silicon Valley: Semiconductor Chip Production and Clean Room Work

Semiconductor chips are the brains of electronics components and a host of consumer goods today. Without them, our radios, cars, watches, computers, airplanes, refrigerators, clocks, and cell phones would not operate properly. Semiconductor chips are central to the functioning of the global economy. For this reason we refer to chip production and related occupations as the core of Silicon Valley's high-tech economy. These chips come at a dear price for the people who manufacture them.

Although infinite variations exist in the way they are made, it is possible to offer a general description of how semiconductors are manufactured in the "fabs" (fabrication plants):

> The manufacture of semiconductor components starts by growing long cylindrical crystals from molten silicon, the basic raw material, and sawing them into thin slices called wafers. They are then baked in "diffusion furnaces," and the heated wafers are treated with a variety of gases to

> improve their conductivity. Next, a photochemical process is used to create complex circuitry on hundreds of tiny squares, or "chips," on the wafers. After testing and sorting for defective wafers, the final product is typically shipped overseas for the more labor-intensive assembly processes. The wafers are split into individual chips, each with the identical integrated circuit pattern on its surface, bonded to microscopic wire leads, and encapsulated in protective casings. The assembled integrated circuits are then returned to the United States for testing and finishing. The semiconductors and other components are next soldered into printed circuit boards, which are assembled into every kind of electronic device, from computers to elevators.[1]

The areas inside electronics firms where computer chips are manufactured are generally referred to as "clean rooms." These rooms are "clean" because they are kept meticulously free of small particulate matter—which can damage chips—by means of air filtration systems. But the people who perform these tasks are placed at a level of risk that no individual with means to do otherwise would willingly accept. Here we examine the working lives of electronics employees in the core and consider the impacts on their health. These stories are derived from personal interviews, extensive legal depositions from ill and dying workers, and archival and newspaper sources.

People's Stories: Work, Toxics, and Health

Sarah Carpelli has worked in the semiconductor industry since the 1960s.[2] In an interview with the authors she takes note of the importance of historical context, in that during the 1960s, there was no widespread public knowledge of the high-tech industry's myriad hazards:

> I had a neighbor introduce me into the world of technology. This was in the beginning, when the technology first started out. We started in warehouses that were dark and dreary and smelled of chemicals. This was in the late 1960s, early 1970s. I didn't pay a lot of attention to the smells and the chemicals or think much about what it was doing to me personally, health-wise. As the years progressed and I saw some of the things, I thought, we shouldn't be smelling these fumes, ingesting these fumes. We shouldn't see the tires rotting off of the cars when there was

> a chemical spill in the parking lot. I had numerous jobs in the industry. I ran wafers through chemicals. I did chemical testing. I worked on the computers. A lot of it, though, was in the chemical area, in what they called a "clean room." We still had many, many chemicals to work with, sulfate, sulfur, acid, to clean the wafers at the many different stages in the process.[3]

Armanda Esperanza, a Chicana worker at a major electronics firm, had an experience similar to Carpelli's regarding the general lack of knowledge of hazards. Like many elderly Chicanas in Santa Clara County today, Esperanza began her career working in the canneries: "I worked in a cannery for like three months. Right out of high school, around 1965. I would sort peaches."[4] She had no idea if any chemicals were used in those operations (in fact many toxins were in use—see chapter 3). After the cannery closed down in 1968, Esperanza began work at a large electronics firm "in the clean room, and we were measuring parts." She would regularly be exposed to epoxy, a hazardous material. "We didn't think it was dangerous, so we didn't really rush to the bathroom every time we got it on our skin to wash it off, you know." This historical ignorance of hazards may have cost Armanda her health. She now suffers from cancer and is a party to a class action lawsuit against her former employer.

Esperanza and other workers wore "finger cots" as protective clothing. These cots were coverings that only protected the tips of a worker's fingers, despite the hazardous nature of the job. She would use chemicals like methylene chloride, epoxy, and solder every day. Often she wore latex gloves, which would "start to disintegrate" from contact with the methylene chloride and "we would take our gloves off because it was much easier to work with your hands than it was with the gloves, and we would just pull out the parts with our hands."[5] This type of risky behavior in hazardous workplaces is not uncommon.[6] It often stems from inadequate health and safety training, a lack of regulation of the workplace by management and state agencies, and the extreme discomfort and lack of productivity (and consequent impairment of one's ability to feed one's family) associated with wearing certain protective equipment. Like many women in the industry, Esperanza was also directly exposed to glycol ethers (a hazardous solvent): "we used glycol. It was like a lubricant, it was a liquid. It was a little abrasive."[7]

A combination of corporate and worker misconduct at Esperanza's firm created further problems, as certain chemicals that had been officially phased out were still unofficially available for use: "I don't recall when they stopped using methylene chloride. It worked great on cleaning heads—so once in a while it was snuck in to clean this and that, but I know they took it off the line completely."[8] She was later diagnosed with breast cancer and underwent a lumpectomy, chemotherapy, and radiation therapy.

Sarah Carpelli recalled the gendered division of labor at her firm:

> Most of the workers were single moms. A lot of single moms. I do remember one woman who had been married and had a child under two years of age. She became pregnant again and continued to work, but had a stillbirth. I wondered at the time, I still wonder if her work had anything to do with that. She worked while she was pregnant. How sad to have to work while you were pregnant.[9]

When she became ill, it was clear that Carpelli's cancer was probably not hereditary:

> After I retired, I was diagnosed with cancer. There was not a drop of cancer in my family in any way, shape, or form, so this was definitely out of the ordinary. My mom lived to be 75, and she passed away due to diabetes. My dad lived to be 99, and he passed away of old age.[10]

Flora Chu is a legal advocate for Asian workers in Silicon Valley. She places the story of one worker, Erlinda Carreon, in a broader political context:

> More than one hundred years ago, they lowered Asian men in baskets to insert and light the dynamite so that mountains could be blasted away and the Sierra railroads built. A hundred years later, Asian immigrants are still asked to put their health and life on the line so that California can prosper. Only now, the job hazards are sugar-coated so that workers do not know the hazards that they are facing until it is too late. Erlinda Carreon was a teacher in the Philippines until she emigrated here so that her daughter could have a better education. Like so many immigrants, she came to the Silicon Valley and worked in the electronics

> industry assembling discs that go into our computers. She put up with headaches, nausea and discomfort for the sake of a paycheck. What she did not realize was that it was the chemicals in her job that prevented her from having another child that she so desperately wanted. What she did not realize was that the chemicals that she worked with were building a cancer inside her. She died of a thymoma, a rare cancer that she realized too late was caused by her work. Erlinda's story is not unique for Asian immigrants. Many Asians form an underclass that works in the low-paying high-hazard jobs under constant threat that they might lose their meager paycheck. They are constantly exposed to chemicals that can permanently disable them. Employers hire Asians into these jobs because they perceive that Asians are a docile workforce willing to perform monotonous repetitive duties without complaints.[11]

Echoing Flora Chu's lament about the treatment of Asians in the Valley, Romi Manan, a Filipino immigrant, discussed the chemical-intensive nature of the work he and other immigrants performed at National Semiconductor Corporation in San Jose. Many of his coworkers made the supreme sacrifice for our high-tech economy:

> There was too much exposure to chemicals. I handled all kinds of chemicals used in the wafer production. At National there were so many issues in regards to health and safety. We tried to address them. Just by looking at the National Safety Papers for every chemical that was used in the plant, you see all the symptoms in other workers. We had so many cases of miscarriage. Glyco-ether is heavily used in the production of wafer materials. We had workers who contracted cancer. It's not easy to pinpoint the cause, but we had a good number of workers who worked around those chemicals who contracted that kind of disease. You would think twice why that would happen. One of the workers who I used to work with died of cancer just some months ago.[12]

Sadly, when workers suffer from occupational illnesses, it affects their whole lives, on and off the job. Furthermore, the impact is felt by coworkers, friends, family, and loved ones. Therefore, we must consider the broader social effects of an individual's struggle to survive work-related health problems.

Disruption of Daily Life and the Family Impacts of Occupational Illness

Paulina Bustamante became so sensitive to chemicals after working for years at a chip plant that it interfered with her daily life outside the job site:

> I'm still experiencing the effects from my xylene [a hazardous solvent] inhalation. I'm real sensitive to certain smells such as chocolate, nail polish, shampoos, and highway fumes. Sometimes in the grocery store when I pass the soap and detergent aisles, I start to feel sick and dizzy again and have to go out to my car and put my head down before I can finish shopping and drive home.[13]

Bustamante's experience was not out of the ordinary. A report on another chip worker stated, "One worker who now reacts to everything from laundry detergent to copying machines as a result of chronic exposure to chemicals on the job, says the job has made her 'allergic to the twentieth century.'"[14]

But the ripple effects of toxic work stretched beyond these survivors' bodies and touched their families in a number of ways. In the most basic manner, these multiple impacts often occur when entire families and social networks are employed in the same firm or industry. Lydia Johnson, a cancer survivor, had deep familial connections to her plant. Similar to many of the older, basic manufacturing industries like iron, steel, and auto, entire neighborhoods, friends, and families often migrated together into workplaces. Silicon Valley at the turn of the twenty-first century is no different for Lydia Johnson, who stated, "My sister, brother-in-law, my brother-in-law's brother, and many close friends from school and the neighborhood I grew up in [worked at the same firm]."[15] Her sister later developed cancer.

As with all basic industries, in electronics these familial connections were advantageous to the industry, serving as a form of social control over potential troublemakers. As labor union activist Michael Eisenscher explained:

> When there was a worker shortage, they relied upon family members [to recruit new workers]. But they also relied upon family members to maintain discipline. If I get my job through you, my cousin, and I am a

> troublemaker, the boss goes to you and you say, "I got you this job, don't screw it up. Keep your mouth shut and do your work." There was a kind of family network of people obligated to one another. You also had the issue of multiple wage earners from one household working for the same firm. You had great family dependency. People all relied on a single employer for their paycheck. The implication was your whole family income could be wiped out if you got in trouble. They'd not only fire you, they'd fire your wife.[16]

This type of family-mediated social control was a major barrier to Jessica Arvada, a clean room worker who told of her family's lack of support for her and their disbelief in her claim that she suffered from multiple chemical sensitivity:

> My aunt also worked at the company, as the secretary for the vice president. She was like a mother to me. I lived with her and my uncle through high school, and my own aunt warned me that I could be blackballed from the company [if I kept complaining about work-related health problems], and she also didn't believe me and stopped talking to me. I still can't believe the whole thing could encompass and come out and hit the home as much, and destroy as much as it has. Not only is there physical damage, but the emotional is so much. . . . I will probably be divorcing my husband as soon as I'm able to go to work. My husband does not believe that anyone can be allergic to anything. The lack of support throughout all of this has been really hard.[17]

Jessica had been exposed to xylene continuously while working at her firm. Her supervisor told her that xylene was entirely harmless, despite its documented negative human health effects.[18]

Other ill electronics workers were caught in a bind because their own children were also working in electronics or perhaps anticipating entering the industry some day. Alida Hernandez worked at IBM from 1977 to 1991 and was always proud to be a part of the Big Blue family. She washed discs used for computer hard drives and was "exposed to water with the chemicals." She said the fumes were so strong that her head would spin whenever she would enter the plant. Like many clean room workers, Hernandez's sense of taste was dulled: "When you went to eat, you didn't taste nothing." Hernandez was diagnosed with breast cancer in 1993 and underwent a mastectomy. She feels a deep shame about her illness and has not yet told her children, who also work in Silicon Valley. Her granddaughter is unaware of her grandmother's condition, and she graduated from San Jose State University with a degree in computer engineering and is considering job offers from Intel, IBM, and Cisco Systems.[19]

Jose Carbon emigrated from the Philippines in 1975. He has worked in high tech ever since then and has seen a lot of illness and death among his coworkers and friends in the industry. He says, "Not only the person [the ill worker], but the family as a whole is disrupted. The wife, the daughter or son. Economically, also, not only the physical well-being of the person."[20]

Daily routines and special occasions become traumatic experiences for many ill workers and their families. Selena Gonzales suffers from

> problems with my menstrual cycle . . . I always have problems with my arms when I'm cooking, carrying things. I burned my arms recently. I'm always starting a fire or something. Nobody trusts me with knives at home. I can't ride a bike no more. I can hardly cook because I always end up burning myself. I don't do hardly any cooking anymore. I can't run. I have a hard time writing. Sometimes I think of things that are way back when, over 20 years ago. Some days I can't remember what I did the day before. This happens all the time. I can't go to activities like I'm supposed to with my kids because I'm always too darn tired. When I have to go to their meetings for the schools, they call them teacher conferences—sometimes I can't even make those. Sometimes I can't even change my own Kotex [sanitary feminine napkin] by myself. Somebody has to help me.[21]

Like many ill survivors of the toxic workplace, Selena's condition altered her ability to care for herself, to contribute to household work, and to do the routine activities her family engages in—yet another illustration that the concentration of women in the electronics production labor force has negative impacts on the ability of those workers who succumb to illness to perform other forms of gendered labor (e.g., in the home).

Another chip worker, Monica, told of how special family events became nightmares. On one Easter Sunday, she had a severe reaction to the cellophane on her children's Easter baskets. She felt suffocated by the fumes. On another occasion, while on a trip to the Santa Cruz Mountains with her family, Jessica had to jump out of the moving car suddenly—a reaction to the vinyl seating and the car exhaust. She recalled, "I felt I had to be totally insane or had flipped. There's no way smells can kill you. It is impossible!"[22]

As a majority of core production workers in Silicon Valley are women, the question of children's health is often at the forefront of their concerns and illness experiences. Lydia Johnson was a senior assembler at a major chip plant and worked in that position for two years. She was forced by the firm to leave work and take medical disability because of her illness. She now suffers from allergic rhinitis, early menopause, and sterility. "I remember doctors telling me that I was sterile. I 'couldn't have children' is how they put it. The Ob-Gyn mentioned since no one else in my family had become sterile or had an early menopause, it was very unusual. I was crying and shocked."[23] As with many other clean room employees in the Valley, Johnson regularly had chemicals splash onto her skin, including her face. And it happened "pretty often. Daily."[24]

In addition to being poisoned at work, another worker, Ronni Martino, like many women of childbearing age, carried another burden, more terrible than the others: "My son had two tumors removed from his brain. He was born with them . . . and to this day, I have thought many, many times of that, thinking that I have done this to my child."[25] Blaming oneself for toxic exposure children suffer is a common way of coping with such a life-shattering event, particularly given that parents, especially mothers, are socially entrusted with the protection of their children. Debra Dodd, a former AMD (Advanced Micro Devices) employee, suffered from asthma after years on the job. She said she constantly smelled xylene when she worked, during her pregnancy, and that her breast milk turned orange as a result.[26]

For those women who had given birth to healthy children, the prospect of dying before their sons and daughters grew up was terrifying. After years of working at a chip plant, Ronni Martino was diagnosed with a rare cancer that had spread to her lungs, bladder, lymph nodes, bowel, and uterus. She remains strong for herself and her family: "I went through hell. . . . My goal is to get well, to stay well and to stay alive as long as I can for my children's sake."[27]

Many women were also economically vulnerable and afraid to speak up about toxics for fear that they would lose their jobs and jeopardize their family's economic stability. A Chicana electronics worker stated, "When I was at work, I couldn't close my hands because they had blisters and they crack and break open and bleed and swell. Of course I couldn't take off from work or quit because I'm a single mother and I had kids to support."[28]

According to one study, many managers in the chip industry referred to the chips as "babies" and the workers as "mothers" whose job was to tend to them. "Workers' identities defined outside the workplace were brought into the production process by this characterization, and became a mechanism of control over workers. By using the language of mother and infant to describe the relationship between female chip worker and her product, the industry implicitly asked workers to make sacrifices for the welfare of the product, even if this meant jeopardizing their own health for the production of flawless chips."[29]

Electronics workers were caught in many binds. They were compelled to choose between their health or their job, but no matter which choice they made, their personal and family lives would be deeply affected. Clean room workers, particularly women workers, found themselves in complicated scenarios where their own health was tied to the health and well-being of their children and family. The physical and psychological stress these women endured interacted to produce and exacerbate illnesses they developed as a result of high-tech work.

Management Control over Work and Well-Being: Power and Denial

Studies of hazardous workplaces indicate that when the balance of power between management and labor is changed even slightly in favor of the latter, when a union or labor advocacy organization is present, the occupational environment often becomes safer for employees.[30] When workers gain some measure of *autonomy*, they have a greater capacity to secure a safer and healthier job. Worker autonomy has been a central feature of sociological and anthropological studies of work during the last three decades. Several studies have highlighted the enduring conflict between workers and managers over autonomy within the labor process.[31] Throughout the nineteenth and twentieth centuries, this issue manifested itself in the form of struggles over craft versus unskilled production, job design, and production quotas, for example. When worker autonomy declines, we observe both an imbalance in power relations and increases in physical and psychosocial occupational hazards.[32] So, from a health and safety or environmental justice orientation, autonomy and power matter a great deal.

One measure of worker autonomy is the degree of participation employees enjoy in decision making on the job. Data from depositions and our interviews with semiconductor and clean room workers indicate that they enjoyed very little participation on the job and that this antidemocratic structure of decision making produced measurable negative health effects.

One sociological study of the electronics industry concluded that management's control over labor is nothing short of "authoritarian."[33] The obsession among chip makers with spotless clean rooms is legendary. David Raric, a cancer survivor, worked as a clean room operator:

> They [the firm] were very big on cleanliness and contamination, putting the safety shoes in the shoe cleaner when you came in or walking on the little, the sticky mats, wearing your gloves, covering your nose with the little face mask part of the smock; a lot of things towards lessening contamination and particles in the clean room.[34]

Ronni Martino was a female employee of a firm that is being sued for allegedly causing her cancer. Her view of the clean room is highly critical:

> I always thought the bunny suit was to protect the employee. I didn't know they were using chemicals. I know that they were making a product, but I thought it was to protect the employee. The floor [in the clean room] was raised with holes in it, because of the airflow going through. They were very concerned over the product, extremely concerned over the product, how you handled it, don't shake it, don't move it, you know, everything had to be the right temperature. They cared more about the product than anything else.[35]

The clean room hygiene policies and the economic necessity that workers keep up their productivity was a deadly combination. The drive to stay on the line and keep "protective" clothing on prevented Lydia Johnson and others from washing off after chemical splashes and other forms of exposure:

> We couldn't wash off much in there. To get to water, you couldn't do that [for fear of contaminating the chips]. We couldn't take our smocks

> off because of the skin flakes and hair. We had to keep our smocks on in that area. [In order to wipe chemicals off her body after a splash, she would] usually use my smock. No rags were allowed in the clean room because of dust and fibers.[36]

Johnson was later diagnosed with cancer.

To their credit, some firms provided workers with information on toxics, at least occasionally. For example, Hannah Anderson worked at a chip plant during the 1980s and remembered that Material Safety Data Sheets "were readily available to everybody" (an MSDS provides information on work hazards that is required, by law, to be posted in all workplaces). Anderson regularly worked with TCE and used freon and solder—substances that are known poisons. Indeed, it is frightening to learn how careful some of these clean room practices were and yet how deadly the health outcomes continued to be. It would appear that, whether companies are careful or careless or whether they provide "information" to workers does not address the more basic problem of the intense prevalence of toxics in the Silicon Valley workplace.

Management and supervisors in the electronics industry have held considerable and virtually unchallenged power over workers. One example of this power—illuminating a theme in most of our interviews with workers and in most of the depositions we analyzed—was management's claim that, no matter how toxic the workplace might be, it was still very safe. David Raric remembers one such example. And like many other workers, Raric's interaction with management also involved subtle coercion:

> I smelled strong odors at least five or six times a day. I mentioned it to the manager and the department techs. . . . But we were told that it was within acceptable limit levels by what the company had set forth and by their testing, and that it was just literally part of the job as working in the chemical process area. It was your job that had to be done.[37]

During the 1980s, workers at the Advanced Micro Devices (AMD) plant in Sunnyvale, California, reported that there were xylene spills two to three times per week, but that their manager would claim "I don't smell nothing" and refuse to allow workers to leave the fab.[38] These examples reveal management's power to deny that there is a problem at all—thereby defining the situation as safe because they hold authority over workers in the corporate hierarchy. Managers therefore place their employees in extremely dangerous situations from which workers can protect themselves only by leaving their jobs.

Management's response to women workers who raised concerns about toxics in the workplace was gendered and patronizing. In 1983, SCCOSH activists interviewed a woman who worked at an electronics plant. She explained how management used the "mass hysteria" argument to defuse any claims of workplace hazards:

> They called it "mass hysteria" when everybody got sick. I'm not talking about just Fab 6; [Fab] 6 and 7 were adjoining, so whatever happened in 6 happened in 7 also, and whatever happened in 7 happened in 6. "Adjoining" means that the clean air ventilation ran through both Fab areas. Okay? So that the original 25 people that went to the industrial clinic [as a result of a chemical spill] that day came from BOTH Fabs. The incident was classified as "mass hysteria." It was like, "You haven't smelled nothing, there's nothing wrong with you, you just smelled a little something and went off the deep end."[39]

The use of the "mass hysteria" charge has an extensive historical lineage. It is at least as old as the charge that women who dared to leave their husbands or those women who claimed they were raped by a man (particularly if it was her husband) were "hysterical" or "schizophrenic."[40] We find the "mass hysteria" and related "mass psychogenic illness" labels employed by medical authorities at state agencies and by electronics firms and governments in other nations as well.[41] In other words, it is a form of patriarchal control, the use of power exercised by men and male-dominated institutions over women to maintain the status quo.

This type of gendered power has measurable consequences for the health and livelihood of many electronics workers. Nancy Hawkes, a former AMD employee, told reporters about her long years of tribulations on the job:

> Sometimes sitting around [at AMD] with other women at break or lunchtime, the conversation would take odd turns. "A lot of women were having menstrual problems. It amazed me how often the subject came up. Some were going through eight, nine, 10 [feminine] napkins a day—and really soaking them."[42]

When Hawkes pulled the employees out of a clean room after finding a puddle of xylene on the floor, her supervisor told her she had imagined the whole thing and that if she continued to talk about the spill "he'd have my job in two days." She surreptitiously got hold of AMD's Chemical Handling Guide and read about the toxic and lethal effects of the materials she had been working with. "It was like reading that someone had put a contract out on your life," she recalled. Her supervisor (the man who had threatened to "have her job" if she persisted with her efforts) told reporters his feelings about Nancy Hawkes: "She's very emotional. She may be somewhat disturbed. [She has a] Chicken Little, sky-is-falling mentality." Hawkes and many other (former) employees battled life-threatening illnesses during and after working at AMD.

When electronics workers go beyond raising the question of toxics and actually request assistance in determining the level of contamination on the shop floor—thus asserting that a problem exists in the first place—management's response is often swift and harsh. Retaliation by management against workers seeking protection from toxics is very common in high-tech manufacturing. Sadly, as sleek and shiny as the Valley's corporate campuses may be on the outside, the level of fear among workers concerning management retaliation in this industry is not fundamentally different from that observed in basic manufacturing such as auto or steel industries, or in the sweatshops of the garment industry. One male worker recalled:

> I used to be a chemical handler at Signetics. I was 19 years old. Most of my job was pouring chemicals. I used to pour them, load them, clean them up and dump them. My work made me dizzy and nauseated. I couldn't sleep at night. These symptoms were getting worse and worse every day I worked there. The second I walked into the Fab area I'd smell a weird odor and get sick from it. I asked for an exposure check [an air quality check] and I was terminated on the spot.[43]

Paulina Bustamante tells a similar story that may provide insight into why accidents and illness may be underreported in the industry. After experiencing an illness, she was

> placed in a new diffusion area. I was there five days when I was put on warning for absences and tardies. I checked into these warnings. It turns

out they were pinning me down for reporting my illness to the nurse during working hours. They said I should have deducted those times from my card. They used time card discrepancy as an excuse to fire me.[44]

Management harassed and fired Bustamante for seeking medical assistance for health problems that were very likely caused by toxics in the firm. Under this type of pressure and coercion, many workers would likely choose to work while ill rather than seek medical assistance on the job.

Fear of retaliation, firing, and "blacklisting" were common among chip workers. However, workers also engaged in acts of resistance against management. A Filipina woman who worked under a particularly brutal system at a firm remembered:

I told the supervisor, "I am not going back to work because there is something wrong with the machine, and I am not standing there and making myself sick over whatever is leaking from it." But it was very few people who refused work because a majority of the people who worked in there were scared to death of losing their jobs. They wouldn't complain, and a lot of them wouldn't like what was going on and they would come to me, complain to me, thinking I could fix it all. But there are so many of them that are scared. There is a lot of Filipinos working there, and they figure their job is everything and they will take just about anything. They don't get up and leave.[45]

With no unions, most workers subsisted day to day at the mercy of their supervisors and management, who could and did punish them at will.

On the surface, many companies protect their image by practicing public relations with their own workers, in the form of trendy concepts like employee "speak out" programs and "total quality management" (TQM). One firm that is believed to have poisoned many workers had an "Employee Speak Out" Program that officially encouraged workers to communicate any concerns about the job to management. Time and time again workers did this, but management never responded to them in any way. TQM was an equally manipulative management strategy. One worker who experienced TQM deconstructed this model of control:

The company always makes the decisions. Sometimes they got workers involved, like in total quality management. That program would say

> that it would empower you. It's like you were being seen by management as their co-equal. If you don't have a union, you sit down with the management and talk about things like a quality job. In such manner you bring up your suggestions and work on problem-solving issues. You would have the feeling that you are important in the company. You call them by first name. You sit side by side with these managers. It works really well. But it corrupts your mind as a worker. You don't discuss your wages, childcare, affordable housing. You just talk about how to improve production. This means that you are being speeded up to do your job. An article came out in the *San Jose Mercury News* [about toxic conditions in the industry]. Management responded by getting involved in asking you what kind of pizza topping you would like. In some meetings they give out pizzas, picnics. They know how to play the game, manipulating you.[46]

The TQM promise of "equality" is generally a farce, considering the highly secretive decision-making practices at most large chip making corporations.

Corporate Secrecy

Secrecy is a way of life at many electronics firms. Managers and executives claim that secrecy is necessary for protecting trade secrets (i.e., "proprietary information" or "intellectual property") from competitors. Other reasons include the fact that many companies are manufacturing products for the military, and some of the technology is classified by the government. Secrecy presents many problems for workers, however, because (1) employees frequently have no idea what chemicals they are working with because the chemicals are labeled with company-specific code names, not their actual contents; and (2) workers labor in departments that are kept physically and socially isolated from their coworkers, thus making it extremely difficult to organize for expressing basic grievances or for collective bargaining. Under such a scheme, not only are workers divided, but if an employee ever claims that health problems are the result of exposure to chemicals, they will have a difficult time even determining what compounds they worked with in the first place. Secrecy, therefore, benefits the industry owners, shareholders, management, and clients, while acting to the detriment of workers.[47]

Armanda Esperanza spoke of the secrecy at her firm and how it fragmented the labor force:

> At the firm, all your operations, I mean you only knew about what *you* were doing. They didn't give you a whole lot—you didn't get a whole lot of information about what was going on with somebody else's stuff. Everything was so confidential that you only knew about what you were doing. How it related to somebody else's operation, you just could never connect it, because it was a lot of things going on. Everything was so confidential and they would constantly be changing, so if anything got out it would confuse the public or whoever's trying to get the information.[48]

This sort of corporate secrecy allowed management to maintain a greater degree of control over the labor process and any potential organizing efforts on the part of workers. After all, how can you organize if you do not even know who your coworkers are and what they do for a living? Workers at Armanda's firm did have access to a Chemical-Right-to-Know class: "You would get on the computer, they would give you information on how to deal with chemicals, and then at the end they would ask you to take a short quiz, after reading the material." However, she added, "*We never had the real name of the chemical.* Like MC was never called by the real name methylene chloride . . . so some of these . . . I can't really recognize as what type of chemical it really was." Other clean room workers reported using chemicals with names like "Blue no. 5" and "Yellow no. 2" but had no knowledge regarding the standard scientific names for these substances. We see this pattern of secrecy across the board in our interview and deposition data from Silicon Valley workers. The profit motive and secrecy conflict with and violate the workers' federally protected right to know about the chemicals they are handling. And when Esperanza brought a suit against the company for allegedly contributing to her cancer condition, at the legal deposition hearing the corporation's lawyer challenged her lack of knowledge regarding specific chemicals she worked with. The attorney also read statements from physicians implying that the plaintiff was deranged, delusional, and suicidal.

When Lydia Johnson would express concerns about toxics and health issues to management and supervisors at her firm, they often responded, "you're safer here than you are at home. So don't worry about it."[49] "I believed them," she recalled. "The firm would not hurt anybody." Moreover, any request to be reassigned to a safer area would be quashed with the reasoning that "what job you're trained on is where they need you, and that's your job, and that's that." She stated that one of her managers lied about the claim that her workplace was cleaner and safer than the average home because she overheard him privately telling another manager that he "didn't like to go in there [the clean room] where the chemicals were, to be exposed and breathe that stuff."

The secrecy surrounding production at Johnson's firm was also a problem during emergencies and industrial accidents. For example, after a chemical spill at the plant, the workers were evacuated. However, Johnson "wasn't told where it occurred, in my area or not. I never heard about what chemicals had leaked or how much had leaked." In 1998, when a class action suit was filed against IBM in San Jose, California, the principal attorney for the plaintiffs, Amanda Hawes, stated, "It is disturbing that some of these materials are closely guarded trade secrets, which is more important than getting to the bottom of their potential for human harm."[50]

The Medical Community's Responses

Few medical doctors can competently diagnose the condition of chemically sensitive, ill, and dying electronics workers. Medical schools still largely reject even the existence of multiple chemical sensitivity; equally disturbing, few doctors have any training whatsoever in occupational and environmental health.[51] The dominant orientation of the medical profession, and the science supporting it, is that "health behaviors" (diet, exercise, etc.) and genetics are the driving causes of most illnesses.[52]

The reasons for these gross oversights are many, but they are largely rooted in the political economy of medicine, which is characterized by corporate funding, some of which comes from companies largely responsible for (and profiting from) environmental and occupational pollution.[53] Not only do many of the companies that fund medical research profit from the pollution of human bodies and the ecosystem, but many of these same firms also profit from the production and consumption of pharmaceutical treatments (medicines, etc.) for the resultant illnesses. For example, W. R. Grace was a company that produced both toxic chemicals *and* medicines for pharmaceutical use. W. R. Grace was also the company widely believed to be responsible for creating a cancer cluster in Woburn, Massachusetts (as a result of TCA dumping).[54] Likewise, General Electric, an electronics company with a history of producing nuclear power—which has generally been linked to a range of illnesses, including cancer—also profits from its medical information systems business.[55]

The genetic fixation with cancer within the medical community is partly the result of pressure by corporations to shift the blame for this disease from polluters onto individual bodies. For example, Ronni Martino, a worker at a chip plant in San Jose, was diagnosed with a rare cancer after putting in several years on the job. Her company doctor asked her if she would like to participate in "some sort of female study. He told me that they would check to see if you had the gene for breast cancer."[56] With the blinders that are typical of the medical profession, in the face of the scores of cancer patients working in (or retired from) the electronics industry, these doctors are funded largely to find the "genetic causes" of these diseases, and are, in a real sense, blaming the victims for their conditions, while allowing the chemical companies and the Valley industries to continue completely unscathed. Some private or independent physicians claim that their industry counterparts do not or cannot publish or communicate the truth about the electronics industry because management places "velvet handcuffs" on them by pressuring them to maintain the industry's sanitized image. The historical basis for such a charge is clear; there are many documented instances of company doctors routinely ignoring or even lying about occupational illness among a plant's workers because of pressure from management.[57] This "suppression of information makes determining which exposure causes which illness a mission impossible."[58]

Generally, workers can choose from two categories of medical authorities: (1) company doctors, and (2) independent, or private, doctors. A company doctor is a medical official, either on site (i.e., at the plant) or at an off-site office, who has a contract with the firm. These individuals' roles as medical healers whose first charge is to "do no harm" are inherently compromised because they are paid by the very firms that may be making workers ill and they therefore have a financial incentive to underreport such illnesses. Independent or private doctors, while not necessarily employed by a firm, are also hampered by the general lack of occupational and environmental medicinal knowledge that plagues the medical community in the United States. Today, with the rapid spread of managed care and health maintenance organizations (HMOs), *all* physicians in these institutions are pressured to withhold treatment and care unless such treatment is viewed as absolutely necessary. This means that the options for ill workers have decreased dramatically in recent years.

Company Doctors

Ronni Martino spoke of the cryptic, guinea-pig–like experimentation she underwent with her firm's medical testing, and her resistance to it:

> I do remember that they called me into medical at one time and they were checking me for something. And after I left, it did bother me for a long, long time because I thought "how stupid of me to even allow it." But they were giving me a physical and they said it was more routine, and the nurse did a breast exam on me and I just felt so odd afterwards because I thought this is like a really strange thing and I was so stupid to ever allow it to happen. So I just thought that was very strange. And from that point on, when they would call you over to medical saying it's time for your physical or whatever, I would come out and tell them I don't want a breast exam, you know, I will go to my own doctor. I didn't know these people from Adam. I didn't feel comfortable letting this woman—I wouldn't feel comfortable letting their doctor do it—this was not a doctor's facility to me. I could kick myself a dozen times when I think how foolish I was.[59]

When asked, "Were you told what the purpose of that exam was?" Martino replied, "No." When asked, "Did you ever learn what the results of that physical exam were?" she also replied, "No, I did not."[60] It was later revealed that this particular firm was keeping secret records of its employees' health status, including what one workers compensation attorney calls a "corporate mortality file"—a thirty-year-old database of those workers who succumbed to various illnesses such as cancer. The cancer rate among this particular firm's employees was found to be much higher than the rate among the general U.S. population, indicating a high probability that the workplace might be killing the workers.

Dorina Hermosa remembered a similar "guinea-pig" type experience when she was working in the clean room of a Silicon Valley chip maker.

One day she noticed radiation leaking from a machine. Management placed a radiation badge on her smock to measure any potentially harmful levels, but neither management nor the company physicians ever informed her of the results. She was not instructed as to how to read the indicator and wore it for thirty days. When she asked what the results of the tests were, management and company doctors told her "everything was okay."[61]

Thomas Fisher's story provides insight into a worker's realization of the problems with company doctors and the need for independent medical advice and treatment. He and several other electronics workers formed a support group through SCCOSH called Injured Workers United:

> Every injured worker had suffered the indignity of being called a fraud by the company doctor. We are not leeches, welfare cases or bums. A company doctor is a medical whore who sells out to the industry and tells ill people they are not sick, and if they are sick it is not the result of their employer. Honest qualified occupational health doctors are few and far between, almost impossible to locate for a critically ill person who is house bound or worse. We wanted a clinic. The California Industrial Medical Clinic, which was founded and supported by the Semiconductor Industry Association was too much a conflict of interest. *Later, this cartel was proven to be guilty of paying doctors to send people back to work within 24 hours, so that no Cal-OSHA report would be filed. This effectively eliminated 90% of all chemical accidents as far as the records would reveal.* Of course the people were all sick still. Nor did we want a radical left wing clinic, which would be anti-industry. We wanted a non-biased source of quality health care and expertise in the field of chemical injuries. After months of efforts we finally convinced the County Board of Supervisors to initiate the Occupational Health Clinic at Valley Medical Center. The OHC was a good start. The physicians were basically honest and qualified.[62]

Fisher's observations underscore the importance of social movements in creating progressive policy changes in the industry and medical community. Without SCCOSH and its Injured Workers United project, for example, the Valley Medical Center would never have been founded.

The next several stories come from *core* workers who had a range of interactions with *independent doctors*. Their experiences do not offer much hope.

Independent/Private Physicians

As we noted earlier, most doctors have no training in environmental or occupational medicine. Therefore, misdiagnosis is routine for many ill electronics workers who seek the assistance of a physician. One worker reported:

> I am 59 years old and worked for 22 years at Fairchild Semiconductor. I am currently on disability due to long-term exposure to various chemicals in the wafer test and fabrication department. . . . I had migraine headaches, ringing in my ears, constant runny nose, irritated throat and difficulty breathing, dry and cracked skin, and eye irritation. At first, I did not realize it was work-related. When I started to notice a connection with going to work, *doctors assured me it was "normal" for workers to get sick due to "overwork" and "stress."* Eventually, each time I worked around solvents I got so sick that I would vomit, get dizzy, or faint at my table. Something created big holes in the wall and caused a new chair to literally disintegrate within a 3-month period near my workstation. If that can happen to the wall and the chair, what can happen to me and my co-workers in time?[63]

Many workers' stories reveal how entirely unaware the medical community remains with regard to chemical-induced illness:

> At work last year, I had chest pains, dizziness, and shortness of breath. Doctors rushed me to emergency for heart attack. I was 36. They placed tubes in my heart to release any blockages, but found nothing. Doctors couldn't explain the cause. They asked me about my family history but not about chemicals at work. My company said the incident was not work-related and I paid the $600.00 hospital bill out of my own pocket. My sister later discovered that one of the chemicals I was working with [antimony] causes heart attack–like symptoms and even cancer.[64]

What is powerful about these two stories is not only the ignorance of medical officials, but the ability of ordinary persons—the workers—to diagnose the cause of their illness.

Most physicians will often freely speculate or authoritatively draw conclusions about the causes of heart disease, common colds, kidney failure, or any of a number of non-chemical-based health problems. But this is almost never the case with electronics workers' illness. As one Cambodian immigrant woman worker told us, "I went to my doctor and he said that since he doesn't work at my job he can't say whether my job caused my illness."[65] The dozens of workers' cases we examined via interviews or legal depositions revealed that in the majority of instances, doctors never asked workers anything about their jobs, and therefore never hypothesized any links between their health condition and occupational hazards, even when the workers had developed cancer or any of a number of other illnesses that can be caused by chemical exposure.

One female worker suffering from cancer stated, "And the doctor—I had to pull information out of him to get him to volunteer anything [about the cause of her cancer]."[66] Fortunately, some doctors were persuaded or concerned enough to write letters to employers suggesting that workers be reassigned to areas that might not cause "allergic reactions" and other health problems. However, other physicians were adamant in their belief that nothing in the workplace could be at fault. For example, after Lydia Johnson visited a doctor for her allergies, the physician wrote to her employer, "The allergic disease is not caused by the workplace." In another letter, a doctor wrote to the firm that she was a "37-year-old divorced woman who is being referred for evaluation of her psychological problem."[67] This of course is alarmingly similar to the traditional "mass hysteria" charge against women claiming occupational illness. Another worker dealt with "quack" doctors who told her that multiple chemical sensitivity "does not exist"; she diagnosed the problem herself when "I started to feel better when I would be away from work. I would go back, I would get worse. That's the way it was."[68]

In addition to misdiagnosis and ignorance, physicians regularly prescribed useless and/or harmful treatments for patients from electronics firms. Anita Zimmerman was "exposed to chlorine gas every day for a month" at her job at AMD in Sunnyvale. After one incident at the plant forced an evacuation, a doctor at an industrial health clinic told Zimmerman to "gargle with sugar water and take cough syrup." Zimmerman now suffers from a form of bronchial asthma that leaves her chronically short of breath.[69]

Doris Santiago spent her entire career in the electronics industry and has always worked with microchips. She worked in the clean rooms of Raytheon, National Semiconductor, and IBM. She suffers from multiple sclerosis, a relapse of cerebral palsy, lupus, and a stroke. She was given a range of medications for her conditions. Lydia Johnson also worked at Santiago's firm: "I had a respiratory disease, chronic bronchitis, chronic fatigue."[70] Doctors also prescribed a witch's brew of medicines for her allergies, depression, and asthma. Thus, after one industry poisoned these women and made them sick, another industry produced medication for these symptoms, few of which the workers felt improved their condition.

Ill and dying electronics workers suffer from the daily horrors of toxic exposure, only to later confront a medical system that is stubbornly ignorant of, and resistant to, any suggestion that the workplace could be a cause of these health problems. The result: pharmaceutical companies continue to reap handsome profits; workers are misdiagnosed and mistreated, stay sick and die; medical research remains mired in antiquity; and the captains of Silicon Valley industry—the original perpetrators of the systemic poisoning in question—go free.

Conclusion

Core workers in Silicon Valley's electronic industry are a dying breed—literally and figuratively. As we demonstrated with the data above, many of these workers' lives are being snuffed out slowly, as a result of years of chemical exposure in the clean rooms. The number of core workers is also declining because all but a handful of chip making plants have left the Valley and moved offshore to Asia, the Caribbean, Europe, or other states in the United States. Some industry officials have argued that this highly toxic work is less of a problem today because so few chip plants remain. But, of course, that process only shifts the toxic burden onto other communities, rather than reducing the pollution.

Semiconductor companies form only one part of the electronics industry. After they are manufactured, the chips must then be placed into printed circuit (PC) boards that ultimately make up the nervous system of the end product. Many of those end products also require a variety of cables. For example, on a typical desktop computer with a monitor, modem, printer, and zip drive, there might be at seven or eight different cables snaking in and out of each component and into the wall socket. The production of cables, PC boards, and printers goes hand in hand. No computer will operate without its cables and computers are considered incomplete without a printer. The workers in these jobs occupy what we call the "periphery" because they constitute the next major stages in the production of electronics. These jobs are also much lower-paid and lower status. We now consider the trials of workers located on the periphery of the high-tech economy and offer a special look into the world of temporary workers.

7 The Periphery
Expendable People, Dangerous Work

Printed Circuit Board, Cable, and Printer Assembly: Revelations and Resistance

PC boards are the foundation of virtually all electronics devices. They are the platform upon which integrated circuits (chips) and capacitors are mounted, and they create the electrical interconnection between components. PC board and cable assembly companies are notorious for being the "smallest, shoddiest, least accountable and most likely to go out of business."[1] They compete with thousands of other assembly outfits (many of them in people's garages and homes) by cutting costs and pushing workers to produce more and more rapidly, often with exacting brutality. Many workers in the periphery earn lower wages than employees in any other occupation in Silicon Valley, with the exception of food preparation and laundry workers.[2] According to the California Employment Development Department, like semiconductor chip making, circuit board manufacturing jobs at the major firms in Silicon Valley such as Apple, Hewlett-Packard, and Sun Microsystems have been stagnant or declining. The reason for this apparent lack of growth at these original equipment manufacturers (OEMs) is that they have outsourced the production of PC boards and other components to contract manufacturers (CMs). Traditional computer equipment jobs at the OEMs declined from 72,080 to 71,383 between 1992 and 1996, while jobs at the CMs increased from 22,858 to 31,321 during the same period.[3] The earnings of employees at the CMs are 30 percent less than that of OEM workers.

On any given day dozens of job advertisements for electronics positions appear in the "Classifieds" section of the *San Jose Mercury News* and the many newspapers in the Bay Area. In the February 1, 1982, and August 26, 1984, editions of the *Mercury News*, for example, numerous announcements appeared for "Assemblers," "Q.A. [quality assurance] Inspectors," "Semiconductor Operators," "Fab Operators," "Micro-Assemblers," "Thin Film and Wafer Fab Operators," "Test/Wafer Sorters," "Stuffers [printed circuit board assemblers]," "Mil Spec [military specifications]," "Aluminum and Gold Bonders," and positions in "Masking and Diffusion." The ads lure prospective employees with questions and claims like "Do you need a stable work environment with excellent pay?"; "You've worked long and hard to get where you are. Now get paid for it. Lucrative shift differentials. We still Offer STOCK OPTIONS!"; "this is a new and revolutionary technology." The employers range from small operations whose names people would not recognize, to some of the largest, such as Varian Associates and Raytheon Semiconductor.

As with semiconductor plants, the majority of workers in assembly firms are women, people of color, and immigrants.[4] SCCOSH, Asian Immigrant Women Advocates (AIWA), and other groups have stepped up to the challenge of providing services to these populations. SCCOSH's Working Women's Leadership Program (WeLeaP!) focuses on educating and empowering women workers in Silicon Valley's peripheral jobs.[5] The program is a multi-ethnic, multilingual effort to provide participants with technical and practical information about the hazards associated with their jobs and the skills needed to negotiate with management regarding these issues. Classes have been conducted for Korean, Latino, Vietnamese, Cambodian, Filipino, and several other ethnic groups. Workers are encouraged to participate in what SCCOSH calls the Worker Stories Process—wherein participants tell "their story" about who they are, their problems and opportunities at work, and their plans for improving their situation. Raquel Sancho, Program Director for SCCOSH and founder of WeLeaP!, explains how the Worker Story Process began:

> We developed a process for workers to communicate with each other. It's a process where you can draw [on paper], or say something focusing on who you are, where you come from and what are you doing here. We always have that every time we meet. The second part is, what is your story? They never pay attention to their body. [Usually the workers] never talk about headaches, other health symptoms. They just pay attention to their work, their family, their children. They buy lotto in case

> they get lucky. WeLeaP! is the first time that anyone has asked many of these women workers to speak about their lives.[6]

Sancho is a Filipina activist who emigrated to the United States in 1987. She has been active in the Gabriela Network, a Philippines-U.S. women's activist network, and regularly travels to southeast Asia to work with women's groups on social justice issues. The WeLeaP! project attracted and graduated more than two hundred Silicon Valley workers in its first two years of operation. Currently, organizations and communities around the United States and in the Global South are being trained under the WeLeaP! model because of its popularity and effectiveness.

The following is a sampling of worker's stories that have emerged from the WeLeaP! program and related SCCOSH projects. Each of these workers is employed in assembly or a related occupation in the periphery.

Celia Chavez, a Chicana assembler in an electronics plant, tells her story about toxic work: "I place wires into printed circuit boards to keep the chips in place. I use a solder gun to bond wires and parts. The gun is hot and it burns off fumes from the flux. The health effect to me: respiratory and allergies. It's hard to breathe. The workplace has no fan or table vent to exhaust away the irritating fumes. I wonder what the long-term health effects to workers like me will be?"[7] Celia took action by joining the WeLeaP! program and seeking advice from SCCOSH about how to change her situation at work.

A young Cambodian woman, also a WeLeaP! graduate, told her story of immigration, hazardous work, and asserting herself on the job:

> I was born in Cambodia and moved to the U.S. at age 7. We escaped from our native country to find a better way of life. We escaped the war in Cambodia to go to Thailand. Anywhere but Cambodia was safe at that time. About 3 and-a-half years ago I worked in computers, doing assembly. That was my second job, right after high school. I worked with chemicals, ultraviolet ink, isopropyl alcohol. We only wore smocks, so our faces were all exposed. I was concerned about my health because the longer I worked, I got more headaches, skin rashes, and dizziness. I went to my supervisor and they didn't do much. I talked to my co-workers who felt the same way [that I did] but they never brought it up, out of fear of losing their jobs. My supervisor said they would do something about the problem, but they never did. I talked to the safety committee

> and they said the company already meets the standards, but I asked, "why do I still feel these pains?" And they didn't answer. My mother tells me to keep quiet, but I'm not like that—I speak out.[8]

This worker is an immigrant—a refugee, in fact—a young female with a high school education. She works with toxics and her major medium for addressing workplace concerns is a management-controlled "safety committee" whose job is to reassure employees that all is well, rather than to promote real change.

In addition to the widespread exposure to toxics, many health and safety hazards in Silicon Valley's assembly firms involve *physical injuries*. Unfortunately, like chemical illness, some of these injuries are "invisible" and difficult to diagnose. Jung Sun Park's story is one such case. She is also a WeLeaP! graduate.

Jung Sun Park emigrated from Korea in 1982. She has a bachelor's degree in human psychology. She worked for ten years as a "stopping inspector" for PC board assembly at two different companies. As an inspector, Jung Sun must stand in one position all day and visually inspect PC boards as they pass through an assembly line. As each board passes, she must bend at the waist for a close visual inspection. This repetitive motion caused a slipped disk in her neck. She began this tedious work just one month after immigrating to the United States in order to earn money and to make a better home for her family. The first few years in the United States were very difficult, as is the case for many immigrants. Jung Sun received minimum wage for full-time work and regularly worked overtime. "It was very depressing; but what could I do? I felt like a baby—I didn't know the language, I didn't know how things worked here, I could not use my education here. I had no friends and everyone and everything was new."[9]

Like many injured high-tech workers, Jung Sun suffered from the incompetence of the medical community. For three years she went to physical therapists and acupuncturists, but to no avail. "I tried everything," she said. "I couldn't sleep, I couldn't button my own blouse, I couldn't do anything." Jung Sun visited doctors routinely. "The doctor didn't understand what was going on. I even changed doctors, but it was the same."

With no scientific explanation for her problems, Jung Sun began to call into question her role as a parent, as a mother: "I became more and more depressed. I thought I was doing something wrong. I couldn't understand why I was so weak. I felt like I couldn't support my two girls. I felt like a failure as a mom and as a worker." It was not until more than fifteen years after the first onset of her neck pains that Jung Sun finally discovered the cause. A neurologist performed an MRI on her neck and discovered the slipped disk. Given a choice between surgery and exercise, Jung Sun chose the latter.

> I've felt like a handicapped person for so long. I guess I am. But now, it's a matter of mind control. I've come to accept this pain as part of my everyday life. I just can't let it control my life. Now, I am trying to find joy in my life again. I've started to draw and paint now and it makes me happier. I have to find other forms of happiness to deal with this.[10]

After so many years of pain from work-related stress, Park finally found a way to address the problem. Many workers are not so lucky.

WeLeaP! participants are taught, above all, to assert themselves on the job. Seeta is a Filipina immigrant female worker at a tech firm in the Valley. She has two jobs that pay $16 per hour. On one of these jobs she is the Quality Inspector, and she proudly states, "I'm worth $1.5 million a day because, although I only get paid $16 an hour, I make 3,000 pieces a day worth $500 each."[11] She experiences a lot of stress from the job and this impacts her family, particularly her relationship with her daughter. She credits her training with SCCOSH's WeLeaP! program for giving her the confidence and skills to request and receive a substantial pay raise from her boss.

WeLeaP! participants and thousands of Silicon Valley workers were recently given hope when the word spread about a successful case of an assembler seeking justice from his employer. Raj Jayadev, a South Asian American man, went to work for Manpower Inc. and Hewlett-Packard in 1998. Like many other Valley workers, he confronted and resisted toxic work and labor exploitation. Jayadev prevailed, and his story was made public:[12]

> I worked at the start of the line. We would grab half-made printers off pallets delivered by forklifts, place them on a conveyer belt, insert formatters and screws, then send them on. Over and over—each line of 30 had to assemble and pack over 1,000 printers a day. The work required strong hands, quick feet, and a back flexible enough to take all that twisting, bending, and carrying. But it also seemed I was having trouble

breathing. None of the others were surprised to hear this. They told me about nosebleeds, asthma, and similar problems, and said it was part of the job. Then one day, there was a temporary shutdown. To keep us occupied, our supervisor asked me, as the volunteer safety committee rep, to lead a "safety meeting." Usually this meant making sure everyone was wearing their smocks and had their hair tied back to avoid contaminating the products. Instead, I asked what sort of health and safety concerns people had at work. This was a new question, so everyone stood silent a while. Then, from the middle of the sea of blue smocks, Barbara cleared her throat and said, "Yeah, my asthma has been acting up a lot lately, and I keep getting that bronchitis." As heads nodded in empathy, Raquel spoke of similar problems, then the idea caught fire. Over half the line workers spoke emotionally about recurring respiratory illnesses. This naturally led to talk of the bad air quality at work. As safety representative, I asked for an air quality check. The response [from management] was, "No inspection is needed. It's just the season, lots of pollen in the air." No one thought that was true, but we got the same answer every time we asked.[13]

Jayadev reached out to a community organization for help, and suffered retaliation in the process:

Then I found a worker advocacy organization called the Santa Clara Center for Occupational Safety and Health (SCCOSH). They told me every employee has a right to get a "Material Safety Data Sheet" describing every substance we work with. We asked for such a sheet for the chemicals that came off the ink cartridges we inserted in the printers. Management was reluctant but after a couple of weeks we had the information sheet which noted, in very small type, that the ink included "Carbon Black" which has been linked to respiratory irritation and is a "possible carcinogen." We pushed harder for an air quality check. My supervisor informed me that I would no longer be allowed to lead safety meetings, "People got too riled up, that's not what the meetings are for." When I said I would go to OSHA if we did not get the air quality test, my supervisor moved me to another line on the other side of the building. I was "a troublemaker." After a series of meetings with managers, verbal warnings, and a written reprimand, I got a phone call one evening after work from a Manpower representative. She told me my as-

> signment at HP had unexpectedly ended, and that I had to hand in my badges.[14]

Jayadev continued his fight for justice:

> We (me and a co-worker, Eliza) went back to the Santa Clara Center. They offered free assistance to help us file a claim with the State Industrial Relations Board. We filed, but a week later Eliza pulled out because she feared a record of fighting with an employer would hurt her chances for finding work—a risk she could not afford on account of her son. Finally, more than 20 months after the original filing, the California labor commissioner ruled that Manpower had violated my right to express health and safety concerns at work. As a remedy, the commissioner called on Manpower to stop discriminating against employees exercising their health and safety rights, to post a notice at the warehouse of the ruling, purge their files of any reference to my unlawful separation from employment, and give me a week's back pay (around $240). [But] workers at the printer warehouse and others like it are still breathing in the same dangerous substances. It's clear that laws designed to protect workers' health have meaning only if people fight for them. Public regulatory agencies alone cannot be trusted to enforce them in a timely fashion. Worker associations and unions, then, must not only hold employers accountable, but also work to hold state agencies to their word. Unions, worker advocacy organizations, and workers themselves must fight to make health and safety laws real, usable, and accessible tools for employees to protect themselves at work.[15]

Raj Jayadev is like many Silicon Valley workers we spoke to in that he recognizes injustice and exploitation and does what he can to maintain his health. However, Jayadev is an exception to the rule in that he was able to take his struggle public. Few workers are able or willing to take such steps because of the very real likelihood that they will be banned from the industry and their entire family's livelihood threatened. As a single, childless, and well-educated man, Jayadev was conscious of the fact that he had more resources and less at risk. However, he still suffers from respiratory problems incurred during his time at Manpower-HP, and no matter how modest his victory, it still counts.

Cable Assembly in Silicon Valley: A Case Study

The Setting: E-Tech

In the winter and spring of 1999, one of the authors conducted ethnographic field research in a cable assembly firm in Silicon Valley. We collected observational, interview, and documentary data over several weeks by interviewing for a job and actually working at the plant. The firm, E-Tech (a pseudonym), is a small operation that produces semiconductor cables and related parts for manufacturers of computer printers, medical research equipment, microwaves, telecommunications equipment, and military technology.

Like many Silicon Valley firms, E-Tech's workforce has a three-tier wage and status structure, with executives and management at the top, administrative support staff in the middle, and production workers at the bottom. The majority of the production workers during the period studied were temporary employees earning between $6.50 and $7.00 per hour. E-Tech's production workforce totaled fifty persons, all of them immigrants from Mexico, El Salvador, India, China, the Philippines, Vietnam, Laos, and Cambodia. The shop floor was a mélange of cultures, languages, and dialects—all factors that would serve to make formal collective action quite difficult. As in many plants in the area, numerous employees came from the same family or neighborhood. Of the fifty persons, forty were female and ten were male. Many of these immigrant laborers were used to low-status, dangerous work. One Chinese employee informed us, "When I was thirteen years old, I worked in a factory in China, packing and assembling radios. It was very hard work."

Production takes place on the ground floor of this typical Silicon Valley "campus"—a picturesque, sprawling white building with a nice lawn and a row of trees at its borders. Most of the work is done in an assembly clean room. Unlike clean rooms at other Silicon Valley firms that produce semiconductor chips, this room does not require workers to wear "bunny suits" to protect the product from micron-sized dust particles. Instead, this clean room is organized so that semiconductor cables and electrical wiring are used to produce components that meet basic quality assurance standards. In this clean room, dozens of workers sit cramped at tables soldering cables and wires and connecting them to plastic and metal components. The workdays are eight to twelve hours long, depending on how many orders are received each morning.

E-Tech managers enjoyed a great deal of latitude in creating or mitigating hazards in this work environment. Because E-Tech and all other Silicon Valley firms are non-union shops, plant owners and managers make and enforce the rules with little or no formal input from workers. The major decisions that impacted health and safety at E-Tech included (1) the extent of toxic materials use, and (2) replacement of permanent employees with temporary labor. However, workers resisted this system with a variety of tactics.

Toxic Work at E-Tech

Two jobs at E-Tech especially highlight the use of toxics in the labor process: soldering and cable assembly. A dozen workers, mostly women, solder the cables and wires. They sit, leaning over piles of cables at four tables in a section of the clean room. Their job is to clean and solder cables and lead wires in accordance with customer specifications. This requires the use of an epoxy patch system with chemical adhesives. Solderers use a "gun" and a flux to dispense the chemicals onto the cables and wires and fuse them, creating toxic fumes that are supposed to be carried away by a ventilation system. Both solder and flux are documented as significant human health hazards. One product's warning label in this department reads:

> Keep out of the reach of children. Caution: Vapors of heated material may be irritating to eyes and respiratory system. Contains hydrocarbon resin. Precautions: Avoid prolonged breathing of vapors. Provide local exhaust ventilation at process locations where large areas of molten adhesive are exposed. Important Notice: Manufacturer's only obligation shall be to replace such quantity of the product proved to be defective. User shall determine the suitability of the product for his intended use and assume all risk and liability in connection therewith.

After the soldering is complete, the connecting and assembly of semiconductor cables begins. This job is done by two-person teams who bring hundreds of cables to the clean room and spend the entire day attaching male/female connectors to the cables. They use a glue gun containing a toxic resin. At the end of a shift, the box next to their work station contains more than 200 spent tubes of this resin, each displaying a label that warns the resin is a "corrosive" and "hazardous substance." Employees do the work by hand, with no protection (neither face masks nor gloves), in a poorly ventilated room. When one of the authors asked Chang, an employee doing this job, "what's in the gun?" he simply responded, "glue." Another product, "heavy duty silicone," is an integral part of production:

> One day, a lead worker, Ning, told me, "We need to get some Silicone." She walked over to Jay's station, brings it back and sprays a good deal of it inside both ends of the plastic molds Jay had made. This lubricates the mold so that we can slip it onto the cable more easily. She has no hand or facial protection and is getting the substance all over her hand with no effort to minimize contact or wash afterwards. I do the same, because she is demonstrating how to do the job and I feel pressured to comply. After all, "this is the way we do it." I was shocked when I said: "Ning, do you think using this Heavy Duty Silicone like this is safe?" and she responded "Oh yes, it's safe. Don't worry." And she picked it up and began spraying it freely on the cable and putting boots and molds on the cable.[16]

Later, we examined the Heavy Duty Silicone ingredients and warnings. Among other things, it stated, "this product contains chemicals known in the state of California to cause cancer, birth defects, and other reproductive harm. . . . Do not expose to skin." Yet all clean room workers regularly used silicone as a lubricant and exposed their bare hands to it on a daily basis. One worker commented about the general nature of health and safety at E-Tech: "I had no health or OSHA training at all. Although I did have to read and sign a general 'safety first' sheet of paper [an agreement to follow safety rules] which claimed that 'you will be given a health and safety orientation and training on your first day of work.' That never happened."[17]

By law, every employer in the United States is required to provide employees with information on any hazardous substances they are expected to come into contact with on the job. The standard method of employer compliance with regulations is providing a Material Safety Data Sheet (MSDS). In many workplaces these sheets are stored in filing cabinets or posted on walls where workers can read them. However, at E-Tech there was not a single MSDS posted anywhere and no employee we spoke with had ever seen one. At no time while the author was observing this workplace was there any written or verbal communication about the myriad toxic substances workers were exposed to—a violation of federal law. In fact, aside from work gloves, goggles, and an occasional facemask, E-Tech managers provided no protective equipment to employees. These practices are part of a trend in the high-technology industry as a whole.[18]

Recent research on the sociology of work has concluded that psychosocial stress is inversely related to one's position in the occupational hierarchy.[19] In other words, people on the bottom of the ladder experience much more stress than those at the top. Our research findings support this hypothesis, but we also found that exposure to toxics on the job follows the same pattern. That is, newer employees and temporary workers were much more likely to work the dangerous jobs, while senior employees tended to work as "leaders" who supervised entry-level workers or as quality assurance and testing, for example. The quality assurance and testing and stock room workers had been at the firm the longest. They each had a strong relationship with the head supervisor and enjoyed permanent status while the majority of workers were temporary.

Toxic chemicals were embedded in the labor process at E-Tech. Furthermore, the physical hazards associated with toxics use at E-Tech were exacerbated by the social hierarchy and psychosocial hazards linked to the use of temporary labor at the plant.

Temporary Work at E-Tech

During our fieldwork at E-Tech, more than half the plant's production workforce consisted of persons with "full-time temp" status, with no health benefits. As the supervisor explained to the author employed in the plant, "you work here, but you're an employee of another company—the temp service." In fact, under this arrangement, the temporary workers are technically "consumers" of the temporary help firm's services (rather than employees), further complicating their status.[20] This created some concern for workers. For instance, paychecks indicate the source of funds as "NASCO Employment Services" rather than "E-Tech."[21]

The temporary hiring system produced anxieties and a lack of trust and commitment between labor and management, as evidenced by the high turnover rate at the plant. For example, in one three-week period, fully 10 percent of the workforce left and was replaced. As Zhu, a young female line leader, told us, "Mostly everyone is new here. And everybody's a temp. Nobody's full-time."

Aside from the economic concerns, the instability of the employment relationship at E-Tech was perhaps most noticeable in the uncertainty of one's safety and health protection on the job. This was largely the result of a de-linking of employers and employees, through the temporary help system. Consider the following quote from a temporary service information video we observed: "By the nature of our service, we cannot be there to observe and protect you against hazards on the job, so you have to be alert and be prepared to recognize the problems and communicate your concerns to us and the customer."[22] The message is clear: responsibility for health and safety in a temporary work system is shifted from employers to employees. Fortunately, workers opposed these forms of control and exploitation from time to time.

Worker Response and the Culture of Resistance at E-Tech

There were three principal worker responses to shop floor conditions: exit, refusal to use safety equipment, and targeting blatant hazards for mitigation.

The first response to objectionable working conditions was to exit, or quit. Chi, a Chinese-Vietnamese woman, told us, "it's no good there. They don't hardly pay enough and there's no benefits. I'm leaving to find work in Hayward or Fremont where they pay $8.00 per hour." Another worker screamed at her supervisor during the middle of a shift and simply left. While this type of response may be a legitimate act of resistance against a system that dehumanizes workers, the roots of the problem go unchallenged because the workplace they exit remains unaltered. Furthermore, these workers themselves lose twice because they will most likely find another job in the industry that exposes them to equally bad, or worse, toxic and exploitive conditions, perhaps at the bottom of the pay scale again.

The second response to hazards was the seemingly irrational decision not to wear safety equipment. In many cases, even when management provided safety goggles, gloves, or facemasks, workers simply did not use them. Often this was because the equipment was cumbersome or inconvenient, or because workers were in a rush, or because they felt the equipment was unnecessary. Some studies of workers who sometimes fail to protect themselves conclude that male workers in particular often refuse protection as a form of machismo and others do so because they view the threat of hazards as insignificant.[23] The discussion of chip workers in chapter 6 confirms that workers sometimes invoked the latter rationale in the semiconductor clean rooms. By not wearing safety equipment, however, workers potentially exacerbated an already hazardous working situation. Given management's refusal to provide health and safety information, an atmosphere of ignorance about toxics prevailed, so workers' risky behavior was in many ways quite rational.

The third worker response to shop floor conditions at E-Tech was more direct and organized: identifying and targeting blatant hazards in the workplace. One afternoon, a worker in the soldering department—the most hazardous area—informed some of the line leaders that the ventilation system was malfunctioning. It was no longer removing the hazardous fumes from the soldering process and was probably contributing to the coughing fits afflicting several workers. Our field notes read:

> All solderers have to bend over to get a close look at their work. Lian's face was about 5 inches from her soldering gun, and smoke was emanating from it right into her face. Lian's co-worker, Cynthia, has a cough that has been louder and louder in the weeks that I've been here. I note that she doesn't smoke cigarettes. Lian, doesn't smoke either, and also has a constant cough.[24]

Like other, more conventional "clean rooms" in electronics plants, the E-Tech clean room involved the use of plastics, inks, resins, solvents, solder, flux, and epoxies. Also like many other electronics firms, the design of the plant often interfered with the removal of dangerous fumes (i.e., ventilation) generated by this work.[25] In fact, problems with faulty or nonexistent ventilation systems in Silicon Valley workplaces seem to be so widespread that it appears to come with the territory.

Yan, a section leader, showed the author and another worker a drawing and a cardboard prototype of a new ventilation system he was voluntarily designing to address the fumes problem:

> We need a new ventilation system, so here is the design. The problem is that the smoke from the soldering is poisonous and it's all around the room. The ventilation system is not carrying it away and it's very bad for health and safety. It can give you problems with your lungs, like cancer.

> It's cancerous, so I want to finish the design before I leave [he was quitting later that month to find work elsewhere].[26]

Yan was very much aware of the health effects associated with this type of work, despite management's failure to communicate these risks to the employees.

Yan and another leader, Javier, walked into the clean room and inspected the ventilation system. The author pointedly asked Yan about the health and safety issue: "Do you think that the reason why so many people in the front room cough so much is because of the poor ventilation?" He responded "Oh, of course. A lot of people doing soldering get headaches and get sick." Javier then reported, "I can smell it now!" We talked to Lei, a woman soldering wires, and she said "it [the vent system] doesn't work very well, so I use the fan." Javier confided in us: "There's not enough vents for all the workers, so that's another reason why the fumes aren't getting suctioned enough [even when the vents are operating properly]."

The ventilation system was obviously not a top priority at E-Tech; rather than hiring a contractor, management had asked Yan and Javier (two perfectly competent machinists, but with no experience or qualifications to design workplace ventilation systems) to tinker with it during their down time. When the author left the plant three weeks later, management was ordering a new ventilation system to correct the problem.

The example of the ventilation system malfunction and repair reveals that, although Javier and Yan had little formal authority to make sorely needed changes in the work environment, at times they were able to mitigate hazards. As with all entry-level Silicon Valley workers, E-Tech employees had no organizational entity from which to draw strength in this case (i.e., unions), so they had to organize informally or consult management on all health and safety concerns.

Race, Class, Gender Inequalities, and Invisibility

E-Tech was a toxic, low-wage workplace with white men in the positions of ownership and management, and people of color—most of them women and immigrants—at the bottom of the hierarchy. This was a classic case of environmental inequality in the Silicon Valley workplace. Race, class, and gender matter in Santa Clara County just as they always have.

One day when Chi, Xiao, and the author were driving home after work, we were talking about the Berkeley Bowl, a community market in Berkeley, and comparing it to Whole Foods, the large corporate health food chain. We were poking fun at the "rich people who shop at Whole Foods," and Xiao said, "Well, it's always the rich who can afford to be healthy—it's always that way." He carried this critique into the workplace as well, when he told me "this job is very stressful and hard on my body. I hope that I can get the education I want so badly, because without that, I'll always be doing this kind of work." Xiao, like other E-Tech workers, was intensely aware of class inequalities.

Gender mattered a great deal at E-Tech as well. The majority of production employees were women, and women did the most body-cramping, tedious work in the plant, such as soldering and assembly. Men tended to do the cable cutting, plastic mold injection, metal welding, heavy lifting, and packaging boxes of cables. But the unofficial work that women did was also noteworthy. The body language and deference rituals women engaged in when in the presence of management were gendered as well.[27] Once, when Lu drove the author home, the author asked her about the manager (a white man who regularly passes through the clean room, inspecting products). Although she had been at the plant for more than a year, Lu responded, "I don't know his name, I just smile when I see him." Some research on the workplace finds that these gendered behaviors are sometimes more important than female workers' productivity in the eyes of male supervisors.[28]

The race, class, gender, and immigration status of E-Tech workers allowed management to ignore the normal codes of civility and consideration. This was a marginal population, after all, and sometimes they were invisible to management:

> Today several men in business suits were given tours of the plant. They were potential clients and the manager was showing them around. He was bragging that "I've been able to keep my costs lower than my competitors." What amazed me was that none of the employees had been told that this visit was going to happen, so there was no effort to "put on a show" of any kind. This was strange because every other workplace I have been in or read about is characterized by management instructing workers to "keep up appearances" when inspectors or clients come calling. It was as if the employees did not fit into this equation; we were background, scenery, or props. The Chief Executive boasted to

> this audience of clients, "People always ask me how I do it with this company and why we're so successful, and I tell 'em it's one guy—Jason (the manager). That's the only reason I need to give 'em—just one person." He smiles and laughs and somehow forgot about the 50 persons all around him who are actually producing the products! In the meantime, a bizarre scene ensued, with the clients inspecting "the product," poking, touching, tugging, and stretching the wires and cables while totally ignoring the dozens of immigrant men and women all around them. The workers silently and politely stepped aside or shifted their bodies to allow the clients access to their workstations. I felt like we were invisible and treated like cattle or slaves at an auction in there.[29]

Invisibility is a major theme in sociological studies of women of color employed by whites because it is so common in workplaces where, ironically, people of color are unseen but their labor is central to the survival of the enterprise.[30]

These case study data reveal the inner workings of the shop floor culture in a Silicon Valley firm. They also suggest that management wields a great deal of power over workers' health and safety conditions. One's position in the firm was correlated with one's exposure to unhealthy workplace factors such as the toxic chemicals and the temp system. From time to time, however, workers could decide when and to what extent they would be exposed to certain hazards, by choosing when to wear safety equipment, correcting certain hazards, or quitting. Examples like these provide indicators of the potential for E-Tech and other Silicon Valley employees to shape environmental inequalities in the workplace by recourse to a range of informal resistance and legitimate negotiation strategies. However, without a more formal entity such as a union or worker advocacy organization, workplace change will only progress within narrowly defined parameters.

Most of the workers whose stories we present in this chapter are underpaid and face the threat of immediate dismissal if they make waves about health and safety conditions, or if a manager deems them "expendable" or "redundant" during a fiscal downturn or merger. In short, they are all temporary workers in one sense or another. But what of the plight of the thousands of persons officially designated as "temporary workers" in Silicon Valley? And what are the implications of the industry's increasing reliance on this category of worker?

Temporary Work: Small Wages, Big Business

Temporary work is as old as human civilization. It has been used whenever a boss—a farmer or builder, for example—has needed day laborers on a short-term basis. And even after World War II, when Kelly Services and Manpower, Inc. became popular for supplying temps to many firms, "temping" was still viewed as *temporary* in the popular imagination. However, in the last two decades, "temping" has become the norm in many workers' careers, and it has become big business for corporations. In 1984, industry observers were acknowledging that a new category of worker had emerged in Silicon Valley: the "permanent temp."[31] At that time, Santa Clara County had the highest concentration of temporary workers of any area in the nation. The county has held that distinction to this day. Temp work had shifted from the traditional, female-dominated clerical/secretarial jobs—typified by the "Kelly Girl" and "Manpower's Girl in White Gloves"—to a wide variety of occupations. In fact, Hewlett-Packard was able to maintain its public relations promise of having never laid off a single worker by hiring temps for nearly 30 percent of its workforce during the early 1980s.[32]

Temporary firms pour a lot of energy into public relations, given the widespread tendency for workers to prefer permanent employment. Kelly Services forgoes the use of the word "temporary" in favor of the sporty euphemism "Free-Agent": "Between 1980 and 1999 the number of free-agent workers in the U.S. has more than doubled, with 1 in every 5 workers now classified outside the traditional labor force. . . . All of these workers will embrace *free-agency.*"[33] Using the term "free-agent" allows Kelly Services to spin the bad news of the explosive growth in involuntary temporary work as good news. Drawing on another euphemism—"staffing"—the American Staffing Association boasts that it is "creating" jobs in the U.S. economy, for which both firms and workers are allegedly grateful:

> We are the jobs people. The staffing industry matches millions of people to millions of jobs. 2.9 million people per day are employed by staffing companies. 1 million new jobs have been created by staffing companies over the last 5 years. 80% of temporary employees work full time, virtually the same as the rest of the work force. The staffing industry offers flexibility to both employees and companies. People can choose when, where and how they want to work. Companies can get the skills they

> need to keep fully staffed during busy times. 64% of assigned employees say flexible work time is important to them. 81% of companies cite labor force flexibility as the overriding reason for employing contingent and temporary workers. Temporary work provides a bridge to permanent employment. People can try out a prospective employer and showcase their skills for a permanent job.[34]

From a worker's perspective, particularly a production-level worker with minimal education, the above claims are all exaggerations and half-truths. Many temp agencies do offer benefits to employees. However, the reality is that most employees never gain access to these benefits because they must work a specified number of hours over a twelve-month period (often 1,500 or more) in order to qualify—in other words, the highly unlikely scenario of working nearly full-time work in a "flexible" part-time job.[35] In reality, then, temporary work generally means no benefits, no stability, less pay, and more stress. As one observer writes, "[t]he social pact between worker and employer has vanished beneath the rhetoric of corporate efficiency. Temporary work, a sophisticated form of indentured servitude, promises to further distance average working citizens from the American Dream."[36]

It is common knowledge among employers that the use of temporary employees ensures "flexibility" not for workers, but rather for employers. Translation: it is a method of keeping labor costs low and it is a tool of control, specifically for undermining worker resistance (from informal actions to the formal union organizing drives). Employers are increasingly building their own in-house temporary personnel divisions. Employees of such entities are listed as regular (i.e., "permanent") workers and thus are never officially counted as temps. Employers are clear about their needs vis-à-vis temps. As one manager was quoted in a study: "We don't want complainers . . . we don't want people who are going to speak out about wages or working conditions, just people who are confident and willing to work."[37]

Worker-activist Raj Jayadev points out that one of the myths about temporary work is the idea that temps "choose" such unstable employment. "Nothing could be further from the truth; people don't choose these jobs, they just don't have any other options."[38] Studies have documented that nearly all of the growth in part-time, temporary labor in the United States and Silicon Valley is *involuntary*.[39]

Another myth about temporary work is that it is, in fact, temporary. After taking jobs in electronics, many temps in the Valley soon realize that they are "permanent temps"—people whose jobs can be cancelled at any time, but who will remain temps for as long as they work at that particular firm. In other words, rarely is there any hope for transition from temp to full-time (permanent) status. Even the supervisor at E-Tech was forthright about this and openly referred to the workers as "permanent temps." Entire cohorts of new working-class high school graduates, eager to make a living and strike out on their own, are now discovering that they can no longer expect any real stability on a job, no matter how hard they work. As Victor Salcedo, a Chicano worker in Silicon Valley, states:

> A week before graduating high school, me and my friend decided to go to Hewlett Packard and fill out an application for employment. One of my career teachers had said HP was a good place because you could advance quickly. When we arrived at the security post at HP, we asked for Human Resources. Instead, a guard handed us a paper. It was directions to a temp agency called Manpower in another part of town.[40]

In this case, Salcedo was to be employed not by HP, but by Manpower-HP.

In Silicon Valley, temporary agencies employ around 10,000 persons in electronics production. Temporary and contingent employment of all types are more common and growing faster in Silicon Valley than in the nation as a whole. Thus, Silicon Valley provides a snapshot of things to come for the rest of the country, as the high-tech industry becomes synonymous with "growth," visions of "the future," and the use of temporary labor.

A small group of workers does enjoy the "flexibility" of the temp economy, although the vast majority of workers have seen their dreams waste away with the rise of contingent labor:

> For a minority of highly skilled employees, who have learned how to negotiate decent wages for themselves and operate in contingent labor markets, these flexible employment patterns can be beneficial—making it easier to balance work and family responsibilities, and to gain greater control of their own work schedules. But for the majority of both low and high-skilled Silicon Valley residents, the rise in contingent

> employment means increasing economic insecurity, declining wages, little access to benefits and health care, and limited opportunities for advancement.[41]

Between 27 and 40 percent of all employees in Santa Clara County are contingent or temporary workers. This workforce is growing two to four times as fast as overall employment, and virtually all of the county's net job growth during the 1980s and 1990s was attributable to the growth of temporary employment.

Battling the "Temp Slave" Economy

Silicon Valley labor advocates have joined the battle over temp work. The case of Raj Jayadev, the Manpower/Hewlett-Packard worker mentioned earlier, is one example. Much of what Jayadev was doing was raising the issue of corporate accountability and responsibility for temporary workers in the new economy. SCCOSH produced a fact sheet on temporary work that underscored that temporary workers earn less money (36 percent less) than permanent workers; their wages are decreasing (14.7 percent in five years); they are most likely young, female, and of color; and they are part of the fastest-growing workforce in California.[42]

Most of the workers we interviewed for this study were temporary employees. Consider Korina's story. She is a Filipina who immigrated to the United States in 1991. She insists on no audio taping and uses a pseudonym because she is afraid of employer retaliation. She is twenty-four years old and has worked in small assembly shops with PC boards since she arrived here:

> I worked doing PC boards for computers, appliances, TV, radio, machines, starting in 1991. My first job I worked for less than two years and I was laid off. Then I worked for a temp agency packaging disks for computers, for 4 days. Then I got hired in PC board assembly for six months, at a small company in Santa Clara. We do loading the components onto the board, trimming, cutting wires, prepping components, bonding, and taping. We were using chemicals when washing the boards, solder burner made a bad smell. I did inspection, quality control and then got another job through my neighbor at a PC shop. They

> didn't provide us with any glasses or aprons and had no health and safety training. All the jobs I have had are temps. It's a bad thing since you need money, but it's always a short-term thing. It's hard to find a good job. No matter how much you apply, they won't call you or hire you. I made $11.50/hour at my last job. It is not enough money to live.[43]

In spite of the risk, many temporary workers refused to accept their status without resistance. As much as she was afraid of employer retaliation, Korina joined SCCOSH's WeLeaP! program.

Romi Manan works with many temps at his electronics plant. He emphasizes the global nature of the temp industry and the need for workers to confront it directly:

> Temp workers have become permanent temporaries and it's global—it's the same here as it is in my home country, the Philippines. The nations bring in workers for six months and then get a new batch of workers. IBM, Intel, HP, and National Semiconductor all do this. The only way to address this is to fight the companies and the politicians—you have to fight them both.[44]

Given the relative weakness of the U.S. labor movement as compared with European unions, it is no surprise that the temporary labor firms that flourish in the United States are illegal in many European nations. In France, where temp firms are allowed, regulations limit the duration of temporary jobs in order to ensure that temps do not replace permanent employees.

But resistance to the temp economy in the United States has produced significant gains in recent years. The Microsoft Corporation employs approximately 6,000 temps and has had ongoing legal troubles with these employees since the early 1990s, because they have demanded equality with permanent workers. In May 1999, the United States Court of Appeals for the Ninth Circuit in San Francisco ruled that several thousand temporary workers at Microsoft would be allowed to purchase shares of the company's stock at a discount, a benefit they had been denied. "It's now very clear that Microsoft should immediately stop treating its temp agency employees like second-class citizens," said Stephen Strong, an attorney for the plaintiffs. "This is a victory against employers who create two-tiered benefit plans that penalize many of their workers."[45]

Earlier we observed that corporations hire temporary workers not only as a cost-saving method (i.e., lower wages), but also as a way of controlling workers. This is because, legally, temporary workers have never been allowed to organize or join unions. At least that was the case until August 2000, when the National Labor Relations Board handed down a ruling allowing temporary workers to organize into unions and giving them collective bargaining rights.[46] Thus, as exploitive as the temporary services and electronics industries may be, there are reasons for hope.

While printer, cable, and PC board assembly may be difficult, and while temporary status for the thousands of Valley workers might be onerous, at least the workers discussed above were employed under regulated conditions in legitimate Silicon Valley firms. There is a much more degrading, less visible, and more volatile economy in the Valley that is rarely acknowledged—the underground practice of home-based piecework.

Home-based Piecework

Thousands of Filipino, Latino, Vietnamese, Korean, and Cambodian immigrants are working in production jobs in Silicon Valley. They are working out of their homes, not telecommuting like white-collar workers managing web sites and writing HTML programs, but making circuit boards, cables, and other electronics components at piece-rate pay. They labor in their living rooms, bathrooms, bedrooms, and kitchens. Lead, flux, solders, and acids are toxins these workers use in their homes every day—they are the ingredients necessary for the creation of the nervous systems of our electronics, our computers, the infrastructure of the information superhighway. Children, siblings, grandparents, entire families, friends, and neighbors all pitch in to produce these parts, often washing components in acidic solutions in their kitchen sinks. The workers labor sometimes twenty-four hours straight at poverty wages with no benefits, and the ever-looming possibility of dismissal (a reality that all temp workers face). In some cases, home workers are paid as little as a penny for each component produced.[47]

The work done under these conditions generally involves the "stuffing" or assembly of printed circuit (PC) boards or cables. This is some of the most labor-intensive and time-consuming work that occurs in Silicon

Valley, and home assembly is viewed as the most efficient way to do it because, according to one company executive, "it has to be done by hand, and there's no way to speed it up. But the home operation can really jam out a lot in a short time."[48] Thus, "the 'speed' of home assemblers is frequently traceable to long hours, low pay" and the use of entire families (including children and the elderly) who work off the books.[49] Many business leaders involved in home-based operations point out that this type of labor is essentially the same as that which occurs in the garment industry in the Bay Area (and virtually everywhere else in the world), where expensive, name-brand clothing is produced under near-slave-labor conditions.

One electronics firm manager was said to have given a worker several hundred uncompleted circuit boards on a Friday evening and told her, "if these are not completed by Monday morning then you may as well not even clock in when you come back."[50] This particular job, like most, would be impossible to complete without the assistance of several individuals working under the table. Wage, tax, and child labor laws and occupational health and safety regulations are thus routinely ignored and violated. Piecework itself is not necessarily illegal. However, the piece rate is subject to minimum wage and overtime laws. According to the law, companies cannot pay the same employee an hourly wage in the factory and then pay them piece rate for after-hours work. But this is exactly what happens twenty-four hours a day in Silicon Valley.

This may sound like a description of a home-based cottage industry from centuries ago, but this type of labor is now at the cutting edge of production for the hottest commodity in Silicon Valley and the computer industry worldwide: *speed*. In fact, more and more computer manufacturers are using this type of labor because it is cheap and because the turnaround is more rapid than with any other type of non-automated production. Piecework is one of the last steps in the assembly of electronics parts, which are often air-freighted back to Silicon Valley from "export processing zones" in Hong Kong, Malaysia, Indonesia, Singapore, or South Korea, where workers perform assembly tasks on microchips.[51] Workers "stuff" or insert the completed silicon chips and other components into holes in small plastic boards that are at the heart of digital watches, computers, radios, and thousands of other consumer products.

Piecework has been a significant component of many industries since the nineteenth century.[52] And, like the jobs at the "core" of Silicon Valley production, these occupations also tend to feature a disproportionately large contingent of women, particularly Asian and Latino immigrant women.[53] They are often non-English-speaking persons or persons who simply need to supplement the income from their day job with work at home (because their day jobs pay so little in the first place). Women are especially vulnerable to the piecework economy because they can be at home keeping an eye on their children and elders while they labor. Families are deeply involved in this work because the more people pitching in on a job, the more income the family can earn. "Everybody helps," one Vietnamese man states, referring to his wife and seven children. "If they don't have [school] homework to do they help me."[54]

Kiet Anh Huynh, a production manager at Solectron Corporation from 1983 to 1992 and a Vietnamese immigrant, told investigators, "we give the workers 100 [printed circuit] boards and the next day they have to bring back 100 [completed] boards. Maybe at home they do it faster if they have brothers or sisters helping them."

Cuong Tran is a professional pieceworker. He's been assembling cables at home for Wilco Wire Technology, Inc. for ten years, with several siblings. With cables snaking across his living room floor, Tran explains that he routinely works twelve to fifteen hours per day, seven days per week, and frequently twenty-four hours at a stretch on rush jobs. He typically makes about $5 per hour, but can clear $20 per hour if his mother pitches in. His eleven-year-old daughter Mimi and her friends often help out by screwing wires to cable connectors.

Pieceworkers face myriad occupational hazards on the job. One Vietnamese pieceworker, Hoang Nguyen, casually rinses his circuit boards in his kitchen sink and blows them with a hair dryer. He is using lead and flux, which are hazardous industrial materials requiring special handling instructions, none of which are followed in the homes. One female pieceworker told a reporter, "When we soldered at home . . . you breathe all those chemicals. Sometimes we do it all together, the whole family, in the garage." Employees of firms doing home-based work frequently experience neck and back pains, eye strains, sleep deprivation, respiratory disorders, and continuous exposure to toxics.

Some immigrant employees might not want to do piecework, but they are often unfamiliar with U.S. wage and hour laws and have little choice but to do so; others simply feel coerced to do this kind of work despite their awareness of the law.[55] Most home-based workers have regular jobs and are often approached at the end of a shift and asked to do home work.

Home-based piecework is widespread in Silicon Valley and is difficult to combat for three main reasons.

Reason #1: The Rat Race to the Bottom

Silicon Valley subcontractors have an average lifetime of less than two years. Competition is stiff and brutal. In this climate, companies are going to do whatever they have to in order to stay ahead of the game, even if that means working their employees like chattel to make razor-thin profits. So, the piecework game starts when firms hire persons to do home assembly under the designation of "independent contractor," which allows the employer legally to ignore wage, hour, and benefit laws that would apply if the contractor were an employee. Many pieceworkers, however, are not actually independent contractors, and even when they are, they generally employ others who are not—another legal violation. Employers can only afford piecework labor when workers agree to accept very low wages.[56] Equal Rights Advocates (ERA), a women's civil rights organization in the Bay Area, conducted research on piecework in the region. They found that the pay home workers were receiving "did not add up to the minimum wage of $5.75 per hour. One woman [in their study] earned only half that amount."[57]

Most firms who use home-based labor justify their practices because it is believed to be one of the best ways to compete with foreign manufacturers, particularly firms in Asia that can pay their workers much less than the U.S. minimum wage. This rationale taps into the more general sense that all international competition must be bested in the interests of national security and overall economic stability.[58] Under this belief system, business and government are willing to make certain sacrifices, such as humane working conditions. As Joe Razo, a representative of a California State Division of Labor Standards task force, put it in 1980, "the laborers suffer. In home work, the pervasive violation of minimum wage and overtime laws is chronic."[59] Other managers see little wrong with piecework because, as Craig Jorgenson, director of operations for Pulnix America Inc. (in Sunnyvale), stated, "it works for us and it works for them."[60]

This "rat race" also produces a lot of revenue for some of the most powerful companies in the Valley, so regulators might be slow to offend the major beneficiaries of the piecework economy. Several large and mid-sized companies, including Cisco Systems, Sun Microsystems, Hewlett-Packard, Flextronics, and Solectron, have been known to regularly contract work out to home-based assemblers.[61] Solectron makes PC boards for both IBM and Sun Microsystems and has received two Malcolm Baldrige National Quality Awards, one of the most prestigious honors in the industry. The CEO of Solectron (a Japanese American man named Nishimura) was quoted as he received the honor from then President Bill Clinton: "The first time we won the Baldrige award, we came out of nowhere. We were looked on as a dirty, sweatshop kind of industry. That industry doesn't exist anymore."[62] Solectron is the world's largest contract manufacturer and the tenth largest company in Silicon Valley, pulling in $5.3 billion in revenue in 1998.

Reason #2: Immigrant Mobility and Opportunity

The second reason home-based piecework is difficult to eliminate is because many pieceworkers view this form of labor as their entrée into the high-tech economy. However imperfect, home-based assembly opens a window into Silicon Valley and can be a form of entrepreneurship for immigrants with meager resources. Some Asian immigrants have achieved success by working under these conditions themselves and later hiring relatives and friends to do home assembly. Bing Nguyen is the CEO of Bentek, one of the Valley's fastest-growing businesses. He is a Vietnamese immigrant who began as a pieceworker and is now a celebrated high-tech leader.

The piecework economy also provides a way for older, non-English-speaking immigrants to contribute to the household economy when they are otherwise unemployable due to their lack of language skills and the prevalence of age discrimination. One of the authors spent many days applying for low-wage high-tech jobs in temporary employment agencies in early 1999. During this field work, it was strikingly common to see an older immigrant turned away because he or she was unable to take the written employment eligibility exam. One South Asian family we observed was devastated when the temporary agency supervisor informed the grandmother, "I'm sorry but we can't send her out on a job if she can't speak English."[63] For this population, piecework is a ready-made opportunity for making badly needed economic contributions to the family unit.

Reason #3:
Family Ties and Ethnic Networks Are Maintained and Strengthened
Home-based jobs are found through word of mouth in the immigrant communities, and your boss is often your friend or relative. Depending on your perspective, this is intra-ethnic/co-ethnic exploitation or opportunity. However, as one Vietnamese woman—whose family pitches in several nights each week on piecework—explained to a SCCOSH activist, "Why do you all keep saying that home assembly is exploitation? We like it because we work together as a family unit when we do this kind of work, and that's the way we did it back in Vietnam."[64]

These cultural dynamics cast doubt on claims by activists that piecework in particular, and Silicon Valley production work in general, are primarily characterized by exploitation and injustice and by racial and gender discrimination. And it makes the task of remedying these situations even more difficult. Given the real probability of no income or a hazardous job outside the home, home-based piecework can be appealing. However, we maintain that greater safety protections and a livable wage are necessary for workers whether one is employed inside or outside the home.

Avenues for Reform?

Home-based piecework in the Valley originally came to light in 1980, around the time the toxic reality of Silicon Valley was just being exposed. Like the chemical spills and the poisoning of workers, piecework was a shameful blemish on the sleek public relations image the industry's boosters had worked so hard to create. The *San Jose Mercury News* broke the story:

> Beneath the Silicon Valley is an underground of cheap labor in which housewives, aliens, refugees, welfare recipients and others struggling to make ends meet earn less than the minimum wage and do without Social Security and workers' compensation benefits. It is a cash market. The companies that use it are able to eliminate 10 percent of their labor costs that go to payroll deductions, and if employees don't plan on reporting the income, the companies can reduce the pay by that much more.[65]

In response to the *Mercury News* report, the state of California launched an investigation. However, this effort was stopped in its tracks for two reasons. First, there was a lack of government officials who could speak Vietnamese, and many home workers were non-English-speaking Vietnamese. And second, a group of industry leaders met with state labor officials and local politicians to request that they be allowed "to police themselves." Self-regulation has been one of the traditional approaches proposed by the electronics industry (see chapter 9) and has predictably led to continued abuses of labor and environmental laws.[66]

As noted earlier, in June 1999, two decades after the first reports of home-based piecework in Silicon Valley, the *San Jose Mercury News* published an in-depth investigation into these practices, which had grown into a major industry, with Asian and Latino women, men, and their families making up the majority of this underground labor force. Investigators believe there are "several hundred assembly houses" in Silicon Valley, the largest concentration of these outfits anywhere in the nation. The *Mercury News*'s Editorial Board called on the industry to cease this practice, and urged state regulators to enforce the laws, which piecework firms regularly violate. In terms the industry could easily comprehend, the Board wrote: "If they need an incentive: Exploitative conditions are the sort of abuse that invites unionization."[67]

Within a week of the publication of *the Mercury News* report in 1999, the state Department of Labor Standards, the federal Department of Labor, and Cal/OSHA met to launch an investigation of home-based piecework practices.[68] Immediately, many of the companies named in the report ceased sending work home with employees and cut off contracts with full-time home-based workers. In every way, this investigation had an immediate negative impact on the workers, and on the companies' bottom lines. One immigrant pieceworker who was thrown out of work as a result of the report recited a Vietnamese saying: "The boss eats rice, and the workers eat rice gruel. If the boss doesn't eat, we don't eat."[69] In December 1999, the California labor commissioner fined three electronics companies nearly $200,000 for violations associated with home-based piecework. Four other companies have been ordered to pay $284,500 in back wages to workers.[70]

Despite these penalties, there is a need for more aggressive basic legislation regarding high-tech work performed in the home. Activists were dismayed when, in January 2000, the federal Occupational Safety and Health Administration (OSHA) decided to exempt all home offices—including piecework sites—from regulation. One of the major issues was the constitutionality of invasion of privacy and searches in private homes. SCCOSH and the South Bay Central Labor Council wrote a letter to the *Washington Post* protesting this action.[71] And later, at a hearing sponsored by the Senate Industrial Relations Committee in San Jose, SCCOSH argued that more health and employment resources need to be made available to immigrant workers and their families in the Valley, and that original equipment manufacturers must be held legally liable for the working conditions and wages their subcontractors impose upon these employees.[72] SCCOSH also published and distributed its own literature for immigrant families and workers engaged in home-based assembly, which detailed the many hazards involved in this type of labor and provided information on exposure prevention.[73]

Like the garment industry, electronics piecework is organized through a pyramid scheme of contractors, where the original equipment manufacturer outsources the work to "first tier" contractors, who then outsource their work to "second tier" contractors, and so on. What this means is that workers and supervisors in each subsequent tier get paid less and less money, requiring them to work that much harder under unregulated conditions. This pyramid arrangement also allows the original equipment manufacturer to operate with legal immunity because each contractor is responsible for their own working conditions. Activists successfully challenged this liability evasion tactic in the garment industry during the 1990s, as part of a national campaign to bring clothing maker Jessica McClintock to justice for hiring contractors who failed to pay their Asian immigrant women employees. Asian Immigrant Women's Advocates (AIWA) was one of the principal organizing groups that successfully pressured Jessica McClintock to sign a historic agreement that acknowledged original equipment manufacturer liability.

One Cambodian immigrant home-based worker, Kamsung Mao, was interviewed for the 1999 *San Jose Mercury News* investigation. Mao had developed respiratory problems as a result of working with several toxic chemicals in his home while repairing power supplies for the firm Top Line. He later filed suit in federal court against two companies, Top Line and Lite-On, the company that contracted with Top Line. The suit was settled, with Mao winning financial compensation, but the case against Lite-On was dismissed, precluding a precedent-setting ruling for "joint-employer liability" in the electronics industry. However, Mao still made history in that his was the first-ever lawsuit challenging home assembly in the electronics industry.[74] Flora Chu, a veteran Asian immigrant workers' legal advocate in the Valley, argues that "things will not improve unless we change the system. The garment industry made the ultimate owner responsible. That's one step in the right direction."[75]

In our personal discussions with K. Oanh Ha, one of the authors of the *Mercury News* report on home-based work, she indicated that, although the series had received a good deal of publicity and won at least one journalism award, "the prevailing attitude among people in the Valley about piecework is 'it's a free market, so what's the problem?'"[76]

The Prison-Industrial Complex

While the above discussion on piecework takes us to the bottom of the occupational hierarchy in the Valley, unfortunately, some jobs in the electronics industry fall a step or two below that on the status and wage hierarchy. This is the high-tech labor that occurs in *prisons*. To our knowledge, this type of labor is not occurring in Silicon Valley, but we must acknowledge its occurrence elsewhere. It is perhaps the lowest grade of labor in all of high technology.

Lockhart Technologies, Inc. is a circuit board assembly firm that employs prisoners in a Lockhart, Texas, prison to perform this labor:

> LTI, which assembles and repairs circuit boards for companies such as IBM, Dell, and Texas Instruments, got a completely new factory assembly room, built to specifications by prison labor. It pays only $1/year rent and gets a tax abatement from the city to boot. [The owner] closed his assembly plant in Austin, laid off 150 workers, and moved all the equipment to Lockhart, where he pays prisoners minimum wage, as required by federal law. The prison then takes about 80 percent of inmate wages for room and board, victim restitution and other fees. [The owner states], "Normally when you work in the free world you have people call in sick, they have car problems, they have family problems. We don't have that here." He notes that the state covers the cost of workers' compensation and medical care. And, he quips, the inmates "don't go on vacations." One prisoner/worker in this facility reports that he hopes to secure employment in the high tech industry after his release. Unfortunately, in Texas as is the case everywhere else, this industry is more interested in hiring immigrant women than African American men like this inmate.[77]

Many high-tech firms are relocating to Asia to secure the advantages of lax regulation and cheap labor. But given the Lockhart sweetheart deal, it is easy to understand why a company would view this as a more appealing and cost effective alternative to shipping production overseas. This merger between high-tech and the prison-industrial complex reminds and reveals to us the combined power of industry and the state when profits and social control are at stake. Inside these total institutions, individuals who have been convicted of breaking various laws work under conditions that would be deemed appalling and highly illegal outside prison walls. Thus, the long arm of the law stops cold at the prison gates, as at the free trade export processing zones in many Third World nations.

Conclusion

Paralleling the "core" of production work in Silicon Valley, the "periphery" also comprises a disproportionately large number of immigrants and women of color, working for less than a living wage, under temporary labor contracts, without union representation, with few or no fringe benefits, and with frequent exposure to toxics and other hazardous conditions. These workers produce the printed circuit boards that house the semiconductor chips, the cables that serve as the power and communications conduits that allow these products to function, and the printers that complete all desktop computers.

Electronics workers at the periphery of production enjoy few, if any, benefits of the vast wealth enjoyed by the Valley's tycoons. Indeed, these workers are at the bottom of the hierarchy both in the industry and in society. They are invisible, unorganized, fragmented, and isolated from each other; and, in fact, they are both victims and perpetrators of illegal activities in what many scholars have called the "informal economy."[78] The ultimate contradiction involved in these sectors is that the labor these workers perform is at the cutting edge of production, yet they exist at the margins of the industry with no decision-making power to speak of.

We should be clear that, in the face of all of these challenges, these workers still perform their jobs with dignity and derive pride from knowing that their labor is a form of sacrifice for their children and families. They are often able to separate their identity from the job and to find "meaning in demeaning work" despite the difficulties.[79]

The relationship between environmental inequality and the workplace could not be any clearer than it is in Silicon Valley. The major studies of environmental racism focus exclusively on the proximity of locally unwanted land uses (toxic facilities, factories, landfills, etc.) to communities of color, thus neglecting the equally important question of exposure to pollutants on the job. In fact, the small but growing number of studies of workplace environmental inequality present evidence that the poor, immigrants, women, and people of color face occupational exposures at a much greater intensity than those exposures that occur at the community level.[80] This study of Silicon Valley also makes plain the undeniable links between workplace and community-level contamination. From leaking solvents at chip plants that contaminate drinking water supplies, to toxic home-based piecework brought into living rooms from the plants, to the impacts of illness and injury on workers' families, the workplace and community environmental concerns have countless common bonds.

Silicon Valley firms and the high-tech industry have major impacts around the globe. Via economic globalization, the reach of transnational firms has extended beyond the United States and Europe around the planet. Accompanying these economic shifts are many environmental and human health problems that we have observed in Silicon Valley since the beginning of high tech's emergence. Globalization intensifies these problems as many of the dirtiest and most toxic hazards are exported from the United States to the Global South.

8

Beyond Silicon Valley

The Social and Environmental Costs of the Global Microelectronics Industry

Introduction

Near the end of the year 2000, high-tech industry leaders from around the world held a conference in Seattle titled "Creating Digital Dividends." The premise underlying this gathering was the notion that "market drivers" could connect all six billion of the world's people through the e-economy, thereby providing better education, health care, and wealth to even the poorest of communities. Iqbal Z. Quadir, a conference attendee from Bangladesh, claimed, "Poor people are poor because they're stuck in an environment that is stifling. . . . If they are connected, the environment changes and everyone's life gets better." Many high-technology executives, politicians, and global development proponents who benefit from an expanding global economy freely endorse such sentiments. For instance, Bernard Drum, an official at the World Bank in Jakarta who attended the Seattle conference, predicted, "people who do take advantage of those opportunities [from the e-economy] will be ahead of the game."[1]

High technology is often presented as a panacea for many of the world's social ills. Consequently, those who create this technology are our saviors, modern-day knights in shining armor.[2]

Yet, as with Silicon Valley, the electronics industry's globalization has also produced global occupational and environmental hazards. Ironically, immigrants trying to escape harsh and hazardous working conditions in developing nations find themselves in similar situations in the United States. Unfortunately, exposure to hazards, unstable or temporary labor, "homework" or "piecework," and strict gender, race, and class segregation are endemic to the microelectronics industry as a whole. While "globalization" or global capitalism is lauded for breaking down barriers (e.g., trade barriers, nation-state borders, cultural divisions), many other barriers are maintained and erected by this process. For instance, the flow of information regarding human health and environmental impacts of chemical exposure in the high-tech sector is strictly guarded. Often, what little is known in one country is not shared with workers and regulators in other nations. Consequently, transnational corporations (TNCs), under the guise of safeguarding "trade secrets," impose embargoes on information that is imperative for the protection of worker and community health and ecosystems. The globalization of workplace hazards continues to spread as a result of the immense power of these industry giants. Workers' voices are constantly minimized or denied altogether through the industry's tight regulation of information. While many readers might expect this sort of unilateral and dictatorial policy making from firms in Third World nations, we must remember that many of these unsavory practices were first perfected in Silicon Valley, USA.

In addition, TNCs contribute to the increasing inequality that divides workers by race/ethnicity, class, and gender, and they maintain the intense economic dependence of "developing" nations on electronic exports. The structure of this industry serves to maintain First World economic imperialism, rather than fulfilling promises of stronger freedoms or democracies.

Global Reach: The Transnational Electronics Industry

More than nine hundred semiconductor fabrication plants (fabs) are in operation worldwide. Electronics manufacturing has expanded from Silicon Valley and the high-tech corridor along Route 128 near Boston (and a few areas in western Europe and Japan) to the southwestern United States and to nations throughout Asia, Europe, Latin America, and the Caribbean.[3] For example, in 1995, IBM (USA) had plants in 25 different nations and on 5 continents, and reported earnings of $35.6 billion; NEC (Japan) had plants in 26 nations across 5 continents, netting $20.5 billion in revenue; Hewlett-Packard (USA) operated plants in 17 nations and 4 continents and took in $25 billion; Fujitsu (Japan) had facilities in 17 countries and earned $24.4 billion; and Phillips (Netherlands) made computer hardware in 13 countries and raked in $22.4 billion.[4] The continued global expansion of high-tech production, along with initiatives such as the establishment of Free Trade Zones, has had a dramatic impact on the lives of workers worldwide. Human migration is one of the results of this globalization.

Migration is a form of labor mobility sustained by interpersonal networks bridging points of origin and points of destination.[5] Transnational corporations, including the electronics industry, many times function as the necessary "bridge" for immigrants seeking to move from one region to another. Despite the U.S. government's continuing efforts to restrict the flow of immigration inside its borders, the growth of global capital and military interventions and conflicts continue to foster large-scale immigration into the United States.[6] In the United States, the growing prominence of established immigrant enclaves in major urban areas, combined with economic globalization, the decline of manufacturing, and the growth of the service sector, have expanded the supply of low-wage jobs. This, in turn, helps fuel continued migration to the United States.[7]

Economic globalization plays a central role in the experiences of immigrant communities. For instance, globalization frequently uproots people and separates them from traditional modes of existence. In the export manufacturing areas of Asia, for example, the catalyst for the disruption of traditional work structures is the massive recruitment of young women into jobs in the new industrial zones. Once a worker enters this line of work, it is generally a one-way proposition, as traditional economic opportunities in rural areas shrink as a result of depopulation and ecological devastation.[8] It becomes almost impossible for workers to return home if they are laid off or unsuccessful in their job search. For workers in these situations, emigration is one of very few options available.

At the same time that immigration to the United States is often viewed as socially, environmentally, and economically problematic,[9] the migration of U.S. corporations to Third World countries has largely been viewed as part of the natural (and beneficial) evolution of global market forces. In the new global economy barriers to capital mobility are reduced by powerful transnational institutions such as the World Trade Organization (WTO), the International Monetary Fund (IMF), and the World Bank, while borders restricting human migration grow stronger, particularly for those persons considered "unskilled" or poor (who, ironically, make the decision to enter the United States based on dramatic economic, social, and environmental changes in their homeland spurred in large part by policies imposed by TNCs, the WTO, and the IMF).[10]

For many new immigrants, working for a U.S.-based company is nothing new. In fact, for many low-wage workers in Third World countries, it is this experience that spawns the idea of migration to the United States as an option for better opportunities. Ironically, these same companies learn to treat workers of particular ethnicities or nationalities the same no matter where they are located around the world. This is certainly the case for many Filipinos in the Philippines and in the United States. As one worker recalled,

> The attitude of management at National Semiconductor in Santa Clara toward Filipino workers may be influenced by the fact that National also has a plant in the Philippines where it pays workers as little as $4.00 per day, treats them like cattle, and runs the plant like it was a plantation. The reality of the workers in the Philippines is not something that National wants workers in this country to know, but it must influence the views of management here or at least feeds the existing discriminatory attitudes and stereotypes which are held by white supervisors and others in management.[11]

By the early 1980s, Filipinos were one of the largest ethnic groups among Asian workers at National Semiconductor Corporation (NSC) in San Jose, California. This mostly female labor force was appealing to management for a number of reasons: (1) many were fluent in English; (2) many had experience working in Philippine electronics plants; (3) they had a reputation for hard work; and (4) they were considered obedient to authority.[12]

However, conditions at NSC were deplorable. Employees were divided along racial/ethnic and gender categories, paid unequally, and required to work with highly toxic chemicals. As a result, many of NSC's Filipino workers began agitating for collective bargaining rights. In 1981, the United Electrical Workers (UE) union infiltrated the plant and a group of mainly Filipino worker-activists set up the Electronics Committee on Safety and Health (ECOSH). Management's response was swift. Michael Eisenscher, the UE organizer at the time, explained:

> Fearing the rising tide of union support among Filipino workers, National Semiconductor flew its Philippines Human Relations Manager to

> the U.S. to campaign against the union. The message was direct and simple, "Don't rock the boat! You are a lot better off here than in the Philippines." Many immigrant workers supported families in the Philippines with a portion of their U.S. earnings. The suggestion that they might lose their jobs or even be deported was designed to dampen their enthusiasm for the union, and was a message hard to ignore.[13]

In this case, as a corporation operating within and beyond nation-state boundaries, National Semiconductor made good use of its many transnational resources.

For many people, the sacrifice of migration does not necessarily lead to upward mobility, but to a horizontal movement at best. It appears that if U.S. companies overseas offered improved working conditions with livable wages and benefits, there would be little reason for migration. However, such humane provisions would negate the economic rationale for industries to move their manufacturing labor needs overseas in the first place.

Like global capitalism, the advancement of high technology around the world has been accepted as inevitable. Alfonso Molina, the author of a study of the microelectronics revolution, argues that many scholars and business leaders have reified this technology, accepting the myth that it has a life of its own. Molina disagrees: "On the contrary, at every stage it [high tech] has been embedded in and shaped by sociohistorical factors."[14] In part because the promises of high technology are so seductive, many authorities have turned a blind eye to the industry's negative effects. High tech's excesses have also largely gone unchallenged because of the considerable power of these corporations to control information, personnel, and regulators in the interest of profit—as evidenced by the lack of government regulation and the complete absence of unionization among its workforce. When problems or concerns arise, they are generally viewed as "natural" byproducts—growing pains—of an evolution toward a greater good. Molina adds, "Critical social analyses have always been in the minority, and all the more so with the technologies of the microrevolution, which have clearly rekindled hopes and visions of technological utopias."[15]

High-tech enthusiasts have made grandiose claims, presenting the industry as a global panacea for everything from poverty, unemployment, underdevelopment, and immigration to cultural parochialism and gender inequality. However, a closer observation of this industry's transnational impacts reveals the heavy social costs many workers on the global assembly line pay every day. The conditions these workers endure recall a familiar pattern of existence for Third World peoples living under externally imposed systems of domination—First World imperialism.

The Global Expansion of U.S. Firms into Asia

With intense global competition (particularly from Japan) and the increasing costs of manufacturing and producing smaller and smaller microchips, many U.S. electronics firms have either abandoned manufacturing altogether or entered into strategic partnerships with other firms to pool the costs of production. And those firms that continue to construct wafer fabs are building them outside of Silicon Valley.[16] According to Flora Chu, former Director of the Asian Worker's Health Project at the Santa Clara Center for Occupational Safety and Health (SCCOSH), less than a handful of fabs are left in the Valley:

> It's not what it used to be. As far as I can see, for the industry, the reason to stay in Silicon Valley was because the infrastructure was here. No matter how expensive, the equipment manufacturer was across the street, parts and so on. So it made sense to have everything here. Now they've built the infrastructure overseas and the American companies are selling out. The Taiwanese are taking over the basic structure.[17]

This shift has resulted in the export to other nations of the environmental and occupational health problems that plagued Silicon Valley.

Indeed, Silicon Valley is no longer the dominant force in chip making.[18] Taiwanese firms are now a major player in that sector, a dramatic difference from a short time ago (1995), when Taiwan produced just 3 percent of the world's microchips. In 1999, up to twenty new "megafabs" were slated for construction in HsinChu alone, a small city in northern Taiwan.[19]

Like Singapore, South Korea, and Hong Kong, Taiwan was accustomed to the presence of U.S. and Japanese electronics firms because it had played host to a number of their labor-intensive assembly plants in the early stages of semiconductor technology development.[20] In fact, by the early 1980s, 85 to 90 percent of U.S. semiconductor assembly was being done offshore, mostly in Asia. These foreign assembly plants were designed for quick profits with minimal capital investment. In fact, these plants earned at least 15 to 20 percent annual rates of return with a full payoff on their investment within five years: "Offshore labour-intensive assembly operations enabled some TNCs to extract profits quickly and directly in the form of surplus value (profits generated by labour), bypassing the need to invest more capital for the development of expensive labour-saving automated equipment which could replace higher-wage labour at home."[21]

In addition, several U.S. firms built similar operations in a number of Asian countries in order to reduce the impact of potential disruptions in production at any single plant. Romi Manan, a Silicon Valley worker-activist, remembers his experience with National Semiconductor: "Over the years, that's how they managed to distribute their assembly lines into all these different states and countries. That's the way they're doing it. I remember the company director said, 'We don't keep all our eggs in one basket.' The way he said it was, 'if there's a problem here, we don't go broke for that. You still have to keep going.'"[22]

Despite this enormous volume of overseas investment, U.S. firms remained cautious. Given the tenuous political and economic climate in many of these newly industrialized countries (NICs), the more expensive capital-intensive operations were kept at home to ensure continued production and to minimize investment risk. At home, in the United States, there was the certainty of having greater political stability, the necessary infrastructure, raw materials, manufacturing equipment, and skilled personnel.

By the late 1970s, the semiconductor chip industry had matured from vending a novelty product to mass-producing a global commodity. At this time, large transnational corporations began to usurp smaller semiconductor transnationals through mergers and acquisitions. These large transnational corporations also refined their global production strategies. They began to seek even cheaper forms of labor in other Asian countries and invested more capital in these regions:

> Fairchild, for instance, moved its chip bonding operations to countries such as Indonesia and the Philippines, where labour was cheapest, and began conducting operations requiring more capital and skilled personnel—such as testing—in Taiwan, Hong Kong and Singapore. The

> company also began using satellite communications to link a computer at its U.S. headquarters with its offshore plants, placing Fairchild's entire international production scheme under unified management.[23]

"Unified management" is a euphemism for the United States's imposition of its own cultural standards on workers in Asian nations with little consideration of local values and mores.

Also during this time, Japanese chip makers began to rise as strong global competitors. Until the 1970s, the six major semiconductor producers in Japan (Hitachi, Toshiba, Matsushita Electric, Nippon Electric, Mitsubishi Electric, and Kobe Kogyo [now part of Fujitsu]) relied for their production needs on subcontracted, low-wage assembly operations in local suburbs and villages in their own country. But when, in the 1970s, the United States and Europe set trade tariffs and quotas against exports from Japanese electronics firms, the latter began offshore production.[24] In this way, Japanese firms could export to the United States from their plants in Taiwan and avoid quotas and tariffs, since the trade would be registered as originating in Taiwan.[25]

Market Volatility and Broken Promises

Fearful of Japanese competition, U.S. firms responded by seeking cheaper labor and lax trade regulations in many Asian countries. Transportation costs were minimal, given the increasingly small size of the chips.[26] However, these offshore assembly sites were greatly affected by the volatility of the semiconductor market, which routinely experiences periods of heavy demand followed by over-production and demand stagnation. Assembly workers in Asia and Mexico, who enjoy little to no job security, are perhaps the hardest hit by these market cycles. Weak demand and rising inventory levels are therefore blamed for the frequent job cuts across the globe in the electronics industry.[27]

Despite such chaos, many NICs spend a great deal of revenue to attract electronics investors and to import the equipment and hazardous chemicals necessary for electronics production. As part of a larger export-oriented industrialization program, the presence of foreign high-tech TNCs promised rapid job growth and revenue, substantial foreign exchange earnings, debt reduction, and the transfer of capital, skills, and technology to develop domestic industries. For many political regimes in

Asia, electronics production offers the promise of instant modernization.[28] The rationale was that the steep initial costs would be offset by even greater economic and social benefits in the near future. So far, offshore production has failed to meet the expectations of Third World nations.[29]

Industry proponents may argue that many semiconductor companies, fabrication plants, and assembly firms are Asian-owned and are positive indicators of domestic electronics industry development. However, while industrialists in some countries, particularly those in the "regional core" (South Korea, Hong Kong, Singapore, and Taiwan), may have enjoyed significant financial gains, the vast majority of workers and residents in these same countries did not. By and large, the economic benefits were unevenly distributed and highly skewed toward the benefit of a few. Much of the capital generated by labor did not accumulate locally, nor was it reinvested into the domestic economy. And those domestic industries and consumer markets that emerged were developed in ways that suited the transnational corporations (TNCs) and international banks rather than the working people of these countries.[30]

Asian workers in these offshore electronics plants find themselves in a difficult position. Given these nations' intense dependence on the electronics industry, workers view their jobs as necessary, despite the structural inequalities that accompany them. In light of this lack of options, their hope that continued participation in export-based economic growth will lead to poverty reduction is certainly understandable. The growing unemployment, lack of sustained domestic economic growth, and increased monopolization of power are viewed, to a certain extent, as the temporary and necessary costs that must be paid to secure a better future for all. Also on this list of "necessary" costs is the toll on the environment and on human health.

Environmental and Social Costs

A nation's environmental and occupational regulatory regimes play an important role in determining its degree of appeal for offshore production. Specifically, the more laissez-faire the controls on hazardous substances, the more attractive the nation as a site for manufacturing. This is particularly the case for the electronics industry because high-tech production is a chemical-intensive endeavor, and any regulation that hinders the use, transportation, or disposal of hazardous chemicals may be considered a burden the industry would prefer to avoid. This observation is supported by economic studies of various industries that conclude that laws designed to control pollution tend to raise manufacturing costs and lower productivity.[31] The economists who authored these studies suggest that the most effective and efficient way to improve environmental laws is through policies that "harness" market forces.

Such suggestions place a greater priority on "the market" than on people, workers, or the environment. Following this logic, environmental and occupational regulations that are designed to protect the health and well-being of humans must be fashioned primarily around the needs of the industry. Moreover, regulations can only go so far—their limits must be imposed by the needs of the market. Needless to say, many environmentalists and workers find this ideology extremely troubling. One report on the high-tech sector notes:

> It must be kept in mind that an economy which cuts corners on maintaining the environment and the health of its people is ultimately a weak economy. Whereas, an economy which preserves health is a sound economy. Health has an economic basis, and the degree of health and illness in almost any community is largely a measure of the fairness of the political, economic and social forces at work, and a measure of the appropriateness of the industries and technologies we develop and use. Modernization and the use of new technologies mean little to those who become ill, unemployed, impoverished, or oppressed as a result.[32]

Here, centering the needs of workers and their communities provides an alternative definition of a healthy economy. It is conceivable that, rather than hindering technological "progress," addressing health problems in the electronics industry may in fact improve not only the quality of work, but also the form and quality of industrialization itself.

Currently, the seemingly endless supply of cheap labor and few protections on behalf of labor have generated minimal consideration for the quality of workplace conditions. What little attention firms and governments around the globe pay to workers appears generally to focus on how to keep labor costs down. Countries that import hazardous substances and environmentally risky industries have not taken the opportunity to learn from the mistakes of the industrialized nations to avoid the myriad ecological and human health problems associated with the electronics industry. Of course, this is easier said than done, considering the economic and political vulnerability of many Third World nations and their dependence upon the First World–driven global economy.

Learning from earlier mistakes is also difficult given that high tech's occupational and environmental dangers were never adequately addressed in the United States in the first place, prior to their migration to other countries (which have few if any environmental protection laws and little enforcement capability when the laws do exist).[33] Frequently, when toxics are discovered in U.S. high-tech firms, suppression of data and evidence and retaliation against whistleblowers have been the order of the day. Government officials have been barred from conducting investigations and health studies of firms in several instances, even when a proven public health threat exists. As a result, there are to date a limited number of studies on the health of semiconductor workers in the United States and elsewhere (see chapter 5). Furthermore, when unions and environmentalist have publicly challenged industry with proposals to reduce labor exploitation and toxic threats, the response has been to export the most hazardous operations to the Third World. Occupational health and safety expert Joseph LaDou states:

> It can be argued that developed [*sic*] countries such as the United States and Japan are passing their environmentally damaging industries to the NICs, and along with them the long-term costs of environmental remediation which, by the American experience, are quite significant. In effect, to provide near-term jobs and industrial development, developing countries are assuming long-term costs that they are poorly equipped to meet.[34]

The potential occupational health and wider ecological impacts of these transnational environmental inequalities are grave. The global electronics industry workforce is estimated at 2.4 million.[35] What little we do know from epidemiological studies reveals an alarming increase in spontaneous abortions among clean room manufacturing workers globally. However, no definitive study has identified its cause, and other health concerns, such as occupational cancer, have yet to be studied.

As a result, offshore electronics facilities have created long-term stresses on both the surrounding natural environment and human health. Worker advocates in the United States and Asia concerned with high tech have documented increased spending on medical care to treat illnesses caused by chemicals and radiation; the channeling of funds away from social programs and into military expansion by various governments; increased government spending on toxic pollution cleanup; increases in legal activities regarding claims by victims of toxic poisoning; a rise in personal grief, hardship, and suffering by those poisoned; and a sense of growing public unrest.[36]

Ecological Disruptions

Occupational and environmental health concerns have been at the center of public outrage against the electronics industry in Asia. For instance, the recent discovery of arsenic poisoning in oyster beds near the wastewater discharge of semiconductor manufacturing corporations in Taiwan has heightened people's level of concern. This was only the latest in a long list of complaints from people living near high-tech plants, including concerns over air, water, and noise pollution.[37] Unlike the United States, Taiwan has no laws regulating industry development's proximity to residential areas. In many U.S. cities, there must be a one-thousand-foot buffer between industry and residential housing;[38] in Taiwan, residents live just a few feet from the factories. The destruction of oyster beds—a significant source of income for many workers—reveals the negative multiplier effect high tech can have on *other* industries and economic sectors, beyond the immediate ecological disruptions.

There are indicators of positive change in Taiwan. For example, Taiwan is experiencing extreme water shortages. The water scarcity has become so serious that the Taiwanese Environmental Protection Agency has stated that it will soon require new semiconductor fabrication plants to recycle 85 percent of their wastewater—an unheard-of goal in this industry. This technology in Taiwan is more advanced than that used in Silicon Valley. Another sign of hope is the recent transnational alliance between the Silicon Valley Toxics Coalition (SVTC) and the Taiwan Environmental Action Network (TEAN) to confront the environmental injustices emerging from Taiwan's high-tech clusters.

Even so, the general trend in industrial waste management is simply to pass the buck. Malaysia, for example, exports much of its electronics industry waste to the Philippines and Thailand. Other companies are simply stockpiling thousands of drums of hazardous waste, apparently hoping for a miracle, since most of these nations lack waste treatment facilities capable of handling the volume of toxics produced in semiconductor manufacturing. In the meantime, reports abound of illicit dumping of these wastes in rivers, streams, and landfills in many parts of Asia.[39]

These environmental justice issues are, of course, not limited to Asia—they expand around the globe, following the movement of the electronics industry. For instance, similar concerns are raised in Costa Rica, where, in 1986, officials decided to shift the economy from one based on traditional export staples to one centered on computer chips and services. Today, Costa Rica earns more revenue from chip manufacturing than from bananas, coffee, or tourism.[40] With the entry of Intel's semiconductor testing and assembly plants, Costa Rica has become host to the first major high-tech development in Central America; consequently, it is an important site as a model for similar development in Latin America. And, like many Asian countries, Costa Rica still lacks the waste management infrastructure needed to process the massive amounts of hazards this industry creates. According to the Silicon Valley Toxics Coalition, the thousands of pounds of waste generated by the Intel facility in Costa Rica are being transported to a toxic waste facility owned by Romic Environmental Technologies in Chandler, Arizona, near the Gila Indian Reservation.[41] In addition, there are significant environmental concerns regarding the Costa Rica site: (1) the facility sits on the main aquifer for the central valley of Costa Rica, on which more than one million people depend for their drinking water; (2) Intel will not agree to adopt the newest water recycling technology despite the fact that they will use more than one million gallons of water a day; and (3) Intel also will not disclose the specific chemicals, resins, and epoxies being used at the fab.[42]

Environmental Justice Conflicts in the Workplace

The U.S. semiconductor industry, however, contends that it has made significant improvements in workplace safety over the years. Many of these safety concerns were addressed through the introduction of new automation technology that handles many tasks that once required human labor. According to chip manufacturers, their factories met industrial standards of the time and have consistently improved working conditions as they learned about the possible health impacts of various chemical exposures. George Scalise, president of the Semiconductor Industry Association (SIA), maintains that "Using chemicals that have the potential to cause harm doesn't mean it's not a safe industry. It is."[43]

Such declarations by industry proponents have been described as "an institutional presumption of innocence for harmful substances," which results in a "body count"—the trail of death and disease a chemical must leave before it is rigorously controlled in the workplace.[44] Critics of the industry allege that chip makers are well aware that scores of the one thousand separate chemicals and metals used to create 220 billion chips a year are hazardous to humans and that these same companies knowingly expose workers to a wide variety of toxins.[45]

According to one study, in the Philippines, where overseas workers make up more than 4 percent of the total population, at least one "body bag," or coffin, of an overseas worker arrives at the international airport in Manila each day.[46] The Philippine government and various nongovernmental organizations (NGOs) concluded that the vast majority of these deaths appear to be caused by working conditions that include torture, rape, ethnic/racial conflict, toxic chemical exposures, lack of access to medical treatment, severe stress, and hopelessness.

From July 1996 to December 1997, fifty-seven overseas Filipino workers reporting a "mystery illness" were shipped back to the Philippines after they became too sick to work. Forty-six of them were women who worked for Philips Electronics Industries in Taiwan. These women returned scarred, in excruciating pain, and were said to owe exorbitant "recruitment" fees that they had to pay for the "privilege" of working for Philips Taiwan. Five have died of various health complications.[47] Under pressure from workers and their advocates, the Philippine government and health officials determined that the workers' illness was caused by job-related toxic chemical exposure. Apparently, the women developed severe allergic reactions to trichloroethylene (TCE), formaldehyde, and copper sulfate, resulting in lesions on their face, body, and extremities. This "mystery disease," later called the Stevens-Johnson Syndrome (SJS), also attacks the immune system and is therefore potentially fatal. Those women workers who did not die face a long recovery with mounting medical bills.[48] Philips, on the other hand, remains virtually unscathed by these incidents.

Philips (and the high-tech industry in general) benefits from the fact that making a direct link between specific manufacturing materials and employees' illnesses has been extremely difficult. Dr. Joseph LaDou, who has studied the health impacts of the electronics industry for thirty years, explained, "Virtually every known solvent has been used in the industry. But that's the problem. When you have a cauldron of chemicals that's causing cancer and birth defects, it becomes even harder to find a smoking gun. You can't prove cause and effect unless you can narrow down the individual chemicals."[49] One worker advocate described the difficulty of establishing causation in this way: "It's like spitting in the wind." He adds, "A chemical is considered harmless until you get a body count. Whatever happened to the precautionary principle—looking at these things as being dangerous instead of safe?"[50] Currently, the federal Occupational Safety and Health Administration (OSHA) has managed to tighten standards for only 26 of an estimated 650,000 chemicals and chemical mixtures to which U.S. workers are exposed. Most of these chemicals have yet to be studied.[51]

Locating the "smoking gun" is particularly difficult given the fact that an extensive, comprehensive epidemiological study requires the industry's cooperation to supply data on workers' illnesses. Not surprisingly, the industry has refused to do so. The semiconductor companies claim that any data on workers' health, including reported ailments and diagnoses by staff medical personnel, are kept secret for the worker's privacy.[52] However, in our interviews with workers in Silicon Valley, many reported that they were never given the results of the medical exams performed on them by corporate personnel.[53] This secrecy appears consistent with the industry's overall policy of restricting access to information.

Given such restrictions, high-tech workers and their advocates around the world find themselves in an uphill battle to legitimize their claims of occupational health and safety risks and violations. Today, with accelerating global capitalism, even the "body count" test may not suffice as a catalyst for significant changes in workplace conditions. The difficulty of changing costly production methods or developing safer chemical compounds may simply make it more "cost effective" to move toxic operations to another country where worker protections are minimal.[54]

On a global scale, the high-tech sector is producing a host of environmental disruptions everywhere it locates. A major report by the United Nations Environment Program (UNEP) and the United Nations Industrial Development Organization (UNIDO) detailed many of these problems:

> As the electronics industry has expanded it has become evident that the pervasiveness of the industry and its products has the potential to cause

> considerable environmental damage if not well managed. The discharge of liquid wastes has caused groundwater and soil contamination with solvents and heavy metals. The disposal, in landfills, of hazardous solid and semi-solid wastes containing heavy metals has caused deterioration of water supplies in the surrounding areas as toxic components have been leached out. Gaseous emissions for electronic components manufacturing facilities have been shown to have effects on the health of the population in the vicinity of the factories: cases of respiratory problems, cancers and sterility have been reported.[55]

The electronics industry is a good example of the potential hazards involved in exporting technologies without adequate environment, health, and safety testing. With the current momentum toward economic globalization and the reduction of barriers for capital and technology flows, these problems will likely continue to worsen. The globalization of capital also leads to the globalization of occupational and environmental injustices.[56]

Maintaining Social Control across National Boundaries: Corporate Malfeasance, Community Resistance

Semiconductor corporations form a powerful global entity with few institutional constraints. Presently, the public depends largely upon the corporations' good-faith efforts to regulate themselves. One observer predicts at least another ten years of free market "manufacturing mayhem" before substantial changes are made in this industry with regard to public oversight.[57] Given high tech's institutionalized influence over governments in the United States and abroad, we view this prediction as optimistic.

The restriction of information is one of the key concerns of offshore manufacturing workers, who find themselves completely unaware as to what chemicals they are using and what the possible health effects might be.[58] In Japan—the second leading nation in the global electronics technology race—there were virtually no public records or accounts of health and environmental problems associated with this industry as late as 1984.[59]

Little has changed today, as the extent of the health problems in Japan's electronics industry remains largely unknown. However, the lack of documentation does not necessarily imply a lack of problems. This was made evident recently when Toshiba reported levels of trichlorethylene (TCE)—an industrial cleaning agent that is linked to kidney and liver damage as well as cancer—at 15,600 times the permissible level at a factory in Nagoya in October 1997. Then, in June 1998, Toshiba reported illegally high levels of the same known carcinogen in the ground water beneath four of its domestic factories. Shortly after these revelations made headlines, another consumer electronics giant, Matsushita, reported that tetrachlorethylene—another carcinogen used for cleaning semiconductors—had poisoned the ground water beneath four of its factories in the Osaka area. Matsushita suspects that the ground water beneath 80 of its 112 plants in Japan may be contaminated with harmful compounds. If these harmful chemicals are being spilled at such recklessly high rates in ground water and nearby neighborhoods, then the workers *inside* these plants may be suffering even more intense exposure levels. These shocking disclosures were made by industry after residents of Nosecho, a residential suburb of Osaka, focused their public anger on industrial polluters in Japan. Rather than enduring public humiliation, firms decided to come "clean" about some of their dirty secrets.[60]

These "voluntary" disclosures would not have been so readily offered had the public remained silent and had no mobilization occurred. As we have stated throughout this book, progressive changes in high tech have been realized only when social movements, workers, and communities are organized and demand them. And this must increasingly happen on a transnational scale, as high tech and other industries go global. The Silicon Valley Toxics Coalition has followed this reasoning in their tactics. For example, their International Campaign for Responsible Technology was established in 1991 to increase grassroots participation in national and transnational high-tech policy development. The campaign comprises more than eighty representatives from nations such as the Philippines, Japan, Greece, Hong Kong, Poland, Sri Lanka, Malaysia, Germany, Scotland, Australia, and South Korea. The Santa Clara Center for Occupational Safety and Health (SCCOSH) also recognizes the value of global action. SCCOSH organizer Raquel Sancho states, "SCCOSH is working on developing relations with worker organizations in other countries. Globalization is the trend [in economic growth], so we at SCCOSH need to keep up by going global."[61]

In today's world of transnational corporations, community-based protest—in both host and home countries—is imperative for industry reform. In other words, a coordinated protest across national boundaries, targeted toward a single corporation or industry, may result in a more powerful position for all workers and residents. Without such coordination, workers and residents, particularly those in NICs and Third World countries, find themselves at a severe disadvantage. For instance, disclosure laws in developing countries are much more lenient than those in First World countries. On this subject, a twenty-six-year-old electronics worker in Hong Kong made an observation: "Apart from the 'no smoking' signs, I have seen no warnings posted. . . . The workers on my line are not told the names of the chemicals they use, not to mention the harm they might do."[62]

An incident at a Japanese factory in Hong Kong makes the point:

> In 1982, over a two-week period at an electrical factory making DC motors in Hong Kong, owned by Mabuchi Motors of Japan, 193 women workers were sent to the hospital for treatment after breathing high levels of ozone, phosgene and other gases, which had been released slowly by printing equipment, which used ultraviolet light. A highly toxic cleaning solvent, perchloroethylene (perc), was also used in the process. One worker was in [a] coma for four days, and several suffered permanent lung damage. Thirteen of the victims were pregnant at the time of the incident. According to some of the Mabuchi workers, at least six of those women experienced miscarriages or forced abortions due to fetal death; however local newspapers reported only three such cases.[63]

Following this incident, local labor organizations demanded a list of all the chemicals used in the factory, as well as government factory inspection reports. The management at the Mabuchi plant refused to comply with either demand, citing the importance of protecting trade secrets, and a local government official confirmed that workers have no right to view an inspection report since, by Hong Kong law, the matter is only between the government and company management. The workers' outrage at this Hong Kong plant was compounded by the knowledge that such demands would have been met had the incident occurred on Japanese soil. Under Japanese law, the Mabuchi management would have had to reveal to employees the names of some of the chemical substances used in the manufacturing process, and their hazards, at the time of hiring.[64] This example makes clear the urgent need for transnational organizing against—and regulation of—transnational industries. Otherwise, firms can freely choose when, where, and if they will apply humane operating procedures.

Such glaring incidents of unequal treatment are, of course, not limited to Japanese companies. In Scotland, health officials have been puzzled by the high number of cancers among women under the age of 65. In an area of Scotland known as "Silicon Glen," the cancer death rate for women is three times that of New York City.[65] Women workers at the National Semiconductor Corporation's (NSC) Greenock, Scotland, facility reported a barrage of chemical exposures over a number of years. Many of these women have already died or are ill with cancer and have experienced a number of reproductive health problems, including miscarriages. The women were particularly angry and insulted when they discovered that U.S.-based workers for the same company were warned of the dangers of using chemical compounds like ethylene-based glycol ethers years before they were.[66]

The women allege that the company knew of the health dangers associated with this chemical and simply failed to inform them. The company responded that they did not have solid evidence that glycol ethers were a reproductive hazard until late 1988. According to James Stewart, NSC's director of safety and health in the early 1980s, the company knew of the dangers even before 1988. Stewart reportedly circulated several memos in 1981 and 1982 to company officials warning of the potential reproductive risk associated with glycol ethers. NSC's management reportedly ignored him.[67] The women at NSC in Greenock, like those in other electronics companies, continued to work with glycol until 1993—a decade after the company allegedly had scientific evidence of its causal link to miscarriages. And company officials have denied the women's claims altogether, or called them "exaggerations" derived from "guilt feelings" from losing their babies. Paralleling the Mabuchi incident in Hong Kong, this NSC case underscores the way TNCs can treat groups of workers—in this case women—unequally.

Such tactics, which essentially blame the victim, are common across national boundaries. Management regularly responds by individualizing the problem or identifying the concern as a "woman's" issue. Many examples bear out this pattern. Eye irritations, one of the most common complaints in the electronics industry, have been documented in many countries. Independent studies have documented rapid deterioration of eyesight among workers using microscopes in assembly plants in Malaysia, Philippines, South Korea, and Hong Kong.[68] Studies from

Korea, Philippines, and Hong Kong warned of higher rates of miscarriage among women workers in the high-tech sector as early as the 1970s.[69] In addition, high-tech women workers have experienced rape and other forms of sexual assault at or near their workplace.[70] In the face of all of these reports and studies, companies continue to attribute the above problems to individual misuse of equipment, personal living habits, or ignorance.

Interestingly, researchers have documented a recurring phenomenon in which women workers using microscopes during the night shift at electronics assembly plants have complained of seeing ghosts or evil spirits. In these cases, one or two workers begin screaming and then the whole work area goes into a panic, thereby temporarily halting the production line.[71] These incidents, labeled "mass hysteria," have been documented in Singapore, Malaysia, Indonesia, and the United States. While there may be a number of explanations (including chemical exposure), it is also plausible that these women workers are "voicing" their grievances about working conditions, beyond the reach of management's normal social control repertoire. Management is usually at a loss as to what to do when these "fits" occur.

Corporate Resistance Tactics

The high-tech industry has consistently obstructed any attempt at conducting a large-scale scientific study of the occupational health impacts endured by workers. Such a study is imperative for ensuring a healthy workplace. Typically, management claims that workers' privacy is a major concern, but this claim seems disingenuous coming from corporations who have never asked their employees whether they would like to participate in a health study in the first place. Furthermore, corporations often do not provide health and safety, toxics, or medical exam information, even at the request of injured or ill workers themselves. Timothy Mohn, Intel's director of environmental affairs, provided further evidence of this veil of secrecy when he told a number of government, industry, and environmental officials in the United States during the mid-1990s that providing personnel records for the first large-scale study of cancer and birth-defect rates among chip workers "would be like giving discovery to plaintiffs' lawyers." He added, "I might as well take a gun and shoot myself."[72] Plans for the study were then shelved.

Workers must continue to search for strategies that will lend legitimacy to their claims and lead to corporate accountability. While this remains a major problem in Silicon Valley, workers in Third World countries confront a number of additional barriers. One author writes:

> Microelectronics production workers have little or no information about comparative labour conditions in other locations. Management can tell workers everywhere that they are being measured against anonymous workers somewhere else, who supposedly make fewer demands, and work more productively for lower pay. Most workers have limited means for verifying these claims, which helps management in controlling workers and dampening labour organizing efforts. As a worker in Singapore told me in 1989: "How do we know if it's practical to protest when they tell us, in other countries the workers are grateful just to keep their jobs?"[73]

There is a pattern of worker injuries and abuses across national borders within the electronics industry. As disheartening as it is, this commonality can also form the basis of collective organizing for better workplace standards across the industry as a whole (globally). Given that these women, immigrants, and people of color on opposite sides of the world labor under strikingly similar conditions, assembling the same products, sometimes even for the same transnational firms, it is reasonable to view this as an opportunity to communicate the common concerns of workers worldwide.[74] After all, global environmental justice will only be achieved by improving the quality of life for workers and communities everywhere.

Conclusion: The New Imperialism

In February 2001, financier George Soros cancelled a speech in Bangkok after protesters threatened to pelt him with rotten eggs and fruit. Among these protestors were prominent local businesspeople who blame Westerners like Soros for speculating in their currency and thereby contributing to the collapse of the Thai *baht* in 1997. Since the Asian economic crisis, Soros and others have been denounced as "neocolonial" foreign investors who benefit from Asian economic dependence on the American-led global economy.[75]

A number of key Asian countries that have transformed their economies into export-led regimes over the last twenty years are now deriding the International Monetary Fund (IMF) and related institutions for their tendency to impose strict external controls over nation-states, thus perpetuating prior colonial relations between the Global North and South. Prime ministers and other heads of state in the region are now calling for programs of "economic nationalism." In Thailand, for example, the government is drafting policies to assist local producers in their efforts to compete with imports. They have also raised work permit fees for foreigners, and state agencies are prohibited from hiring non-Thais as consultants. Even more surprisingly, in 1998, Malaysia went so far as to impose controls on its capital markets. Despite the fact that the World Trade Organization (WTO) might eventually declare this sort of state intervention illegal, all of these policies drew broad public support domestically.[76]

Nearly two decades ago, critics warned that, while political relationships between Global South and Global North nations may change as a result of the export-oriented development model, economic relationships would remain basically intact.[77] While most Asian nations achieved political independence from European and Japanese colonial rule, they continue to struggle to institute democratic reforms, to increase economic and technological self-reliance, and to improve working conditions for their citizenry. Students of history should not miss the fact that the World Bank, the IMF, and the General Agreement on Tariffs and Trade (GATT) were established by Global North nations with a long history of colonialism. It was apparent that, despite globalization's promise as the solution for poverty, hunger, and other social ills for the South, the most powerful economic entities with a stake in globalization overtly and covertly destabilized attempts at self-reliance by these nations. Neocolonialism, via corporate and transnational political power, is still very much alive. Martin Khor, former professor of political economy and president of the Third World Network in Penang, Malaysia, states:

> Colonial rule—accompanied by the imposition of new economic systems, new crops, the industrial exploitation of minerals, and participation in the global market (with Third World resources being exported and Western industrial products imported)—changed the social and economic structures of Third World societies. The new structures, consumption styles, and technological systems became so ingrained in Third

World economies that even after the attainment of political independence, the importation of Western values, products, technologies, and capital continued and expanded.[78]

In fact, more and more, under free trade agreements such as the WTO and the Free Trade Area of the Americas (FTAA), these neocolonial arrangements are coming to resemble a contemporary brand of imperialism, as the sovereignty of Global South nations and local cultures and economies are subordinated to the will of more powerful nations and corporations.

The failure of export-oriented industrialization results in large part from its fundamental unsustainability. Not enough of the capital generated by workers in those countries is accumulated locally or recycled into the domestic economy.[79] Instead, transnational corporations strengthen their global markets and global base for cheap raw materials and labor by strengthening traditional colonial economic relationships. The continual sacrifice or postponement of basic democratic reforms and self-reliant domestic industries, the direct vulnerability to sudden fluctuations in the world market, and the fact that domestic industries and markets are developed in ways that suit the TNCs and banks, have all exacted a considerable toll on workers and the general public around the globe. Rather than being lifted out of poverty, social unrest, and national debt, these populations have witnessed their increase.

A Third World nation's "progress" or "modernity" appears narrowly defined as those conditions that create greater dependence on transnational companies. A number of observers have noted the striking similarity between the colonial era and the current "new" era of "postcolonial" global capitalism.[80] One author concludes that the Third World continues to function as an ever-expanding market for First World goods and services, and as a source of cheap labor and raw materials for First World industries. He states, "*Development* is just a new word for what Marxists called *imperialism* and what we can loosely refer to as *colonialism*—a more familiar and less loaded term."[81]

Nonetheless, workers are regularly reminded that a bad job is still better than no job. However, we must ask: how bad must a "bad" job be before action is taken? How high does the "body count" have to go before governments will consider industrial reform? The global electronics industry continues to employ a range of racist and sexist justifications for their exploitive divisions of labor to wash away the industry's responsibility to ensure a clean, safe, and humane workplace.[82] Currently, workplace health and safety remains largely a special privilege for the few. Evidently, immigrants, women, and other people of color are rarely considered worthy of these protections.

Inevitably, unionization efforts (globally and in the United States) must confront the threat that such agitation will simply lead to the industry's movement to more "hospitable business climates."[83] However, it is also the case that corporations frequently pick up and leave with little notice, whether or not workers mobilize, and often in spite of record-breaking profit margins. Management offers little explanation, other than the sacred quest for greater competitiveness. TNCs show no loyalties to workers or local communities and yet expect a great deal from both—not only in tax subsidies but also in hard work, and sometimes workers' (and neighbors') lives. The power imbalances are immense and the working people and residents of the globe's high-tech clusters bear the weight of these environmental injustices.

As much as high-tech firms may operate in different regions of the globe with varying degrees of regulation and oversight, what we find compelling are the striking similarities. The depletion and pollution of human bodies, air, land, and water is endemic and occurs without fail wherever high-tech clusters locate. We began this book with the observation that Spanish imperialism was the earliest form of environmental racism in what is now called Silicon Valley. We are dismayed that the human exploitation and ecocide that accompanied the Spanish conquest are not so distinct from the colonization and imperialism of present-day high-tech corporations establishing market and political dominance around the globe.

9

Toward Environmental and Social Justice in Silicon Valley, USA, and Beyond

Introduction

Since the conquest of the Ohlone people by the Spaniards in Alta California, beginning in 1769, environmental inequalities have been central features of the region's landscape. The destruction of the Ohlone's sustainable culture based on acorn farming and shell fishing gave way to the mission system, based on slavery, the intense degradation of Native women, religious indoctrination, and a ranching economy founded on overproduction and overconsumption of land, water, and animal species. This European system of dominating humans and nature laid the foundation for the next two and a quarter centuries in the Santa Clara Valley.

Like the ranching economy of the missions, the California Gold Rush was a political-economic phenomenon that supported empire-building for elites on many continents. The Santa Clara Valley's abundance of natural resources (i.e., mercury, water) and indigenous, Mexican, and Chinese laborers were the fuel that drove the Gold Rush throughout its short and ecologically unsustainable life.

The region's next system of production, the canneries, was also marked by hazardous—often toxic—gendered labor, first imposed on Chinese men, then later on mostly immigrant women and women of color. The canneries, and the agricultural base that supported them, extracted, polluted, and wasted more water than any previous system of production. And like the Ohlones during the mission era, and the Chinese, Mexicans, and others during the Gold Rush, cannery workers often rebelled against the harsh exploitation visited upon them.

Finally, the Fruit Bowl of America gave way to High-Tech America, as Santa Clara Valley became Silicon Valley. The high-tech economy exploited the region's workers and the natural resource base as much if not more than ever before. Occupational illness, death, overwork, and low wages characterize the jobs of those who assemble computers, cables, and printers; and those who make microchips. Air, water, and land pollution have similarly marred the neighborhoods of Silicon Valley residents, particularly where the working class, immigrants, and people of color reside.

These patterns of environmental injustice spanning more than two centuries years do not instill optimism for the future. However, the continuous resistance by workers and residents does offer some reason for hope. The question we ask is, now what? What does the future hold for Santa Clara County? In the next sections we will consider past, present, and proposed future actions to ensure sustainable and just communities and industries in the Valley.

Environmentally Just and Sustainable Communities and Industries

Forgotten (or Defeated) Historical Sustainable Alternatives

The struggle for environmental justice and sustainability is nothing new. Santa Clara County's history is replete with practices and proposals that would reduce human and natural resource exploitation. For example, the Ohlone people's fishing, farming, and mining practices were sustainable, but they were completely disregarded by the Spanish, who instead preferred over-harvesting, overproduction, and the depletion and despoliation of fisheries, land, and minerals. The Spanish-Mexican use of communal *acequias* was ignored as well by Anglos in San Jose and elsewhere, as corporations began to buy and sell all available fresh water. The criterion for access to water then shifted, from whether or not one was a member of the community (for use toward meeting basic needs) to whether or not one could pay money for the water (with no regard for how much one used or for what purposes).

The use of the *fong sei* by Chinese miners is another example. The *fong sei* was a place near the gold mining camps where Chinese workers would congregate to affirm the value of nature.[1] But all of these ideas and practices were pushed aside as private corporations took over California's economy.

Another failed opportunity to engage in sustainable planning occurred at the dawn of the electronics industry's growth spurt, in the 1950s. Santa Clara County distinguished itself as a pioneer in the establishment of "exclusive agricultural zones" for the purpose of protecting some of the richest farmlands in the area.[2] By the late 1950s and early 1960s, tens of thousands of acres of green space had been set aside for protection. These lands would soon be devoured in San Jose's annexation wars and by the growth of high-tech firms.

More recently, after the Vietnam War, the "Peace Dividend" and "military conversion" (whereby these industries and government funds would be redirected toward the public interest in some way) projects showed great potential. Neither of these endeavors came to fruition, however, as the United States once again prioritized military and nuclear power during the cold war.[3]

The most recent struggles for sustainability involve the promise of pollution prevention and labor empowerment—or environmental justice (EJ). Pollution prevention is an idea promoted by environmentalists and the USEPA and publicly endorsed by most large computer/electronics firms.[4] To be fair, the electronics industry has been successful at reducing many pollutants, if we measure those improvements against units produced. However, environmental initiatives have generally only occurred when grassroots activists or regulators demanded them, and as the volume of production has increased with the growth of markets, the total pollution burden has increased, rendering any per-unit pollution reductions questionable.[5] But the campaigns to ban a single pollutant or class of pollutants will always be limited to an uncomfortable mixture of chemistry and politics and will always fail to target the much deeper problems of power and inequality. A number of decentralized, innovative organizations are pursuing labor empowerment in the Valley. SVTC, SCCOSH, AIWA, and other EJ groups and "worker centers" have headed up the struggle. The AFL-CIO–affiliated South Bay Labor Council, which is closer to the "worker center" model of organizing than to the trade union model, has assisted non-union workers in gaining access to benefits denied them by high-tech firms. Through WeLeaP!, the Justice for Janitors Campaign, the Campaign to Ban TCE, and the anti-piecework campaigns, these groups have led mobilizations to empower women, immigrant, and youth workers in the Valley, to provide collective bargaining rights and health care access to temp workers, and to pursue joint employer liability for a host of wage violations and for the impacts of chemical production and exposure. These actions range from the reformist to the radical and remain the best hope for environmental justice in Santa Clara Valley.

Environmental Justice and Sustainable Community Development

The World Commission on Environment and Development, a United Nations body usually referred to as the Brundtland Commission, popularized the concept of sustainable development. In 1987, the Brundtland Commission published a report in which it defined "sustainable development" as practices that allow societies "to meet the needs of the present without compromising the ability of future generations to meet their own needs."[6] In general, the debates over sustainable development have evolved into two separate camps. One school, exemplified by the Brundtland Commission and most economists and policy makers, centers on the extent to which economic growth can occur while maintaining the viability of ecological systems and their biological diversity.[7] In other words, the integrity of the economy is the first priority, and ecological sustainability is a secondary goal—a sort of bonus, if and when it can be attained. The Brundtland Commission and most other published texts on sustainable development suffer from one particularly deep flaw. They fail to question the growth imperative at the core of capitalism. The scholar Herman Daly defines sustainable development as "development without growth . . . beyond environmental regenerative and absorptive capacity."[8] Daly has led the charge among scholars arguing that sustainable *development*, as used by most scholars and policy makers, is a euphemism for sustainable *growth*, and that the latter simply cannot exist within the Earth's finite ecological limits. He and others argue that all forms of economic growth deplete ecosystems and, therefore, that truly sustainable solutions will only be arrived at by devising ways to meet the economic, social, and biological needs of human beings in ways that are predicated on "steady state" systems of development.[9]

The second school of sustainability proponents includes scholars and activists who use the term "sustainable development" to focus on *community development*. Here the emphasis has been placed on developing projects that achieve economic stability, social justice/equity, and environmental protection—or sustainable community development (SCD).[10]

Activists and scholars working in this area have placed social justice and equity on the sustainable development agenda and have placed environmental concerns on the economic development agenda. This school of thought has been highly critical of post–World War II forms of economic development in the United States and around the globe. During this period, the primary goal of economic development has been to generate handsome profits for corporations, meager wages for workers, and revenue for the state, with little thought of any other social or environmental needs. When concerns are raised about social and/or environmental issues, the solution to these problems is said to be rooted in greater economic growth. Growth is believed to generate the revenues and technological advancements needed to solve social and environmental problems.[11] Proponents of sustainable community development (SCD), including the authors of this book, are seeking to support forms of community development that simultaneously generate economic vitality, environmental protection, and social justice/equity. The core question here is: what are the conditions or practices that will strengthen the local economy of struggling communities while also rebuilding strong social systems and preserving the environment? Not many widespread practices meet these conditions and criteria but a few approximate at least some aspects of this type of development. Sustainable community development would mark a dramatic shift in the future trajectory of capitalism.

Sustainable community development (SCD) and environmental justice (EJ) overlap considerably. We define EJ as a scenario in which no communities are unfairly burdened with environmental toxins and where systems of production are socially and ecologically sustainable and operated under democratic decision-making structures that empower both community members and workers. EJ "is supported by decent paying safe jobs; quality schools and recreation; decent housing and adequate health care. . . . These are communities where both cultural and biological diversity are respected and highly revered and where distributive justice prevails."[12] EJ is not only focused on development or economic systems; it is also about the daily functioning of communities with the highest priority placed on those activities that sustain human and environmental health. EJ must take into account the differential ecological impacts on populations across dimensions of race, gender, and class. It is our contention that sustainable community development must go hand in hand with the goal of environmental justice.

As we have emphasized throughout the book, progressive and radical changes often—if not always—originate from mobilizations among social movements. In the following sections we consider the role and impact of social movements on the high-tech sector.

Social Movements, Advocacy Organizations, and Resistance

We divide the actions of social movements in Silicon Valley into two broad categories—environmental and labor campaigns. The principal points of this discussion are that social movements are the drivers of progressive change in society and that these movements have already had measurable impacts throughout the Valley's history.

Environmental Campaigns

LEGISLATION

Legislation is perhaps the most visible and measurable impact social movements can produce. The Right to Know (RTK) movement of the 1970s and 1980s was led by occupational health advocates and environmentalists who helped pass state and federal legislation that required industry to disclose the volume and nature of pollutants released. While these laws are highly imperfect, they have provided a great deal of ammunition for the environmental justice movement in its campaigns and lawsuits against particularly egregious polluters. Currently, an extension of this earlier movement is pushing for the development of a "Global Chemical Right-to-Know," which focuses on industries worldwide. SVTC and the International Campaign for Responsible Technology—both of which were involved in the previous struggles for national and state RTK legislation—are spearheading this project. These movement groups are aware that, for years, many polluting industries have left regions, states, and nations where environmental laws are more stringent, to locate in "pollution havens" in the southern United States or the Global South. What this means is that the USEPA and many state EPAs can claim reductions in various pollution sources only because they have been exported to other locations, rather than being prevented or reduced. A Global Chemical Right-to-Know system might provide data on how these fleeing industries are operating once they have moved to more laissez-faire zones. This development is also an extension of the call, by one hundred nations at the 1992 Earth Summit in Rio de Janiero, to establish national pollution registers. If implemented, this would be called the United Nations Pollution Release and Transfer Register Protocol. At this time, however, only nations in Asia and Europe have said they would participate. Regardless of which countries sign on, though, the Protocol would allow for citizens to access a range of pollution data.[13] As with the U.S. RTK laws, the major questions will be: (1) how accurate and useful are the data? and (2) will movement groups be able to mobilize around this information and translate this knowledge into power?

In earlier sections of this book, we noted that Santa Clara County has the most federally designated toxic Superfund sites per square mile of any place in the nation. Superfund site listing is as political as any other environmental policy making, and in fact, one reason why the County has so many sites on the register is because social movement groups pressured the EPA to take notice. In 1984, SCCOSH Executive Director Pat Lamborn stated,

> In June and July the EPA said this area's contamination was "insignificant." . . . In our meetings held this summer we demanded that all 120 identified sites in the Valley's groundwater basin be added to the Superfund list. [The public pressure] turned the EPA's head around. Now we're listed as the key county in the country with the most sites going onto the Superfund roles.[14]

SVTC and other environmental organizations, along with groups like the Santa Clara County Fire Chiefs Association, were also responsible for getting two landmark environmental laws passed in the county. The 1983 hazardous materials storage ordinance and the 1989 toxic gas ordinance were major accomplishments because they required firms in the county to go far beyond the environmental standards established by state and federal laws.[15] The passage of these laws revealed that EJ groups were not the only ones challenging the Valley's toxic polluters. The Fire Chiefs Association had been struggling with Valley firms for years because the latter refused to release vital information about the nature of chemicals being used—a serious handicap for firefighters attempting to put out fires on burning high-tech campuses. The 1989 law required firms to classify all gases under various categories of risk, providing crucial information to firefighters about how rapidly certain gases would spread into a community, therefore assisting them in evacuation planning. With the above laws in place, Silicon Valley workers and residents were much safer, and social movement organizations are to be credited for this.

Policy and Practice: Sustainable Developments?

Social movements have been at the forefront of defining concepts and enforcing principles of environmental sustainability in the high-tech industry. One such concept is called "eco-efficiency," a model of business activity that attempts to achieve environmental protection while preserving industrial integrity and profits.[16] Many transnational firms and trade groups claim to be engaged in some sort of eco-efficiency or green practices, including the Semiconductor Industry Association, 3M, Monsanto, Dow Chemical Company, BP, Amoco, and others.[17] For example, the Semiconductor Industry Association boasts that the "U.S. semiconductor industry is one of the most active promoters of safe environments of any manufacturing industry in the world."[18] Sun Microsystems states, "An important part of Sun's commitment to responsibly manage all aspects of its business is the environmental stewardship of its products."[19] And Apple Computer's Environmental, Health and Safety Policy Mission Statement reads,

> Apple Computer is committed to protecting the environment, health, and safety of our employees, customers and the global communities where we work and live. We recognize that by integrating sound environmental, health, and safety management practices into all aspects of our business, we can offer technologically innovative products and services while conserving and enhancing resources for future generations.[20]

However, given that these same corporations are part of larger networks of industries engaged in the wholesale destruction of natural habitat, watersheds, and human health, this is not nearly enough.[21]

A concept related to eco-efficiency is that of Design for Environment (DFE), the idea that manufacturers can become more sustainable when they use less toxic materials in production and design their products so that they can easily be disassembled, recycled, and/or reused at the "end of life." Germany's "Green Dot" program has worked to achieve this goal with auto manufacturers, who are made responsible for collecting and recycling the products after consumers are finished with them. Social movement groups concerned with the electronics/computer industry's growing contribution to the global waste stream seek the same thing. Activists from Silicon Valley joined with environmentalists from around the United States in April 2001 to launch a National Electronics Take it Back Campaign to encourage computer makers to adopt a DFE approach to production.[22] Their concerns arise from the following observations:

- More than 12 million computers are disposed of annually. This statistic is increasing as computer lifespans decrease and as computer purchases and use by consumers rise.
- More than 315 million computers will become obsolete by 2004, creating more than 1 billion pounds of lead waste and millions of pounds of other highly toxic materials.
- More than half of American households have a computer and most users are unaware of proper disposal practices required for computers.

The National Electronics Take it Back Campaign maintains that computers can be cleaned up and safely recycled, "but only if consumers demand it." Their platform is based on the principle of Extended Producer Liability (EPR), which seeks to make brand-name manufacturers and distributors financially responsible for their products when they become obsolete—throughout the product's "lifecycle." "Our ultimate aims are pollution prevention and waste avoidance through a hierarchy of practices, including source reduction, reuse, re-manufacturing and recycling."[23] The platform includes seventeen principles, including reduction of hazardous materials in manufacturing, ceasing hazardous waste exports, ensuring fair labor practices for electronics workers (including fair pay and the right to organize), and shifting the burden of paying for these programs away from taxpayers and back onto the companies.

The National Electronics Take it Back Campaign also promotes the adoption of the *Precautionary Principle*. As they state it, "where there is a threat to health or the environment, a precautionary approach requires taking preventive action even before there is conclusive scientific evidence that harm is occurring. The federal government should develop and implement strict protocols for testing chemicals and mixtures before they are introduced into the markets."[24]

Despite their green rhetoric, transnational high-tech corporations have resisted such environmental responsibility measures, but their success varies by nation, because some regions have a stronger history of environmental and citizen advocacy than others. For example, in response to environmentalists' demands, the European Union now mandates that producers of electronics equipment be financially and physically liable for these products at the end of their consumer life. Similar legislation has passed in Japan, and some U.S. firms have implemented take-back programs in Europe and Asia.

However, some companies have made it clear that they will only comply when legislation forces them to do so, not because of a commitment to some set of higher "green" principles. For example, Apple Computer has take-back/recycling programs in Germany, but not in the United States; Dell has programs in Norway, Germany, Sweden, the Netherlands, and Taiwan, but not in the United States; IBM has operated an electronics take-back in Europe since 1989 and only recently announced a similar program in the United States (the program in the United States requires consumers to pay $29.99 while many European and Asian programs are free of charge); Sony has offered financial incentives for consumers to return old monitors for recycling in Germany since 1996, but its U.S. program offers no such incentives;[25] Hewlett-Packard recently announced a take-back program in the United States, which is also based on a fee for service plan but which is generally viewed as a step in the right direction.[26] All across Europe, in each nation where these take-back programs exist, labor and environmental movements are much more influential and organized than their U.S. counterparts. Thus, industry's variable "commitment" to environmental principles reveals the importance of transnational social movement organizing.

We also note that environmental legislation focused on the computer industry succeeded in the European Union despite industry's lobbying efforts against it. In June 2000, for example, demonstrators piled junked computers and other electronics goods in front of Microsoft Corporation's headquarters in Seattle, Washington. The organizations involved were the Washington Toxics Coalition, Clean Production Action, Asia Pacific Environmental Exchange, Basel Action Network, and the Silicon Valley Toxics Coalition. They were protesting U.S. Trade Representative Charlene Barshevky's lobbying on behalf of the American Electronics Association to stop the European Union's move to pass the Waste from Electrical and Electronic Equipment (WEEE) legislation. The WEEE would require firms to recycle their computer waste and ban hazardous materials in electronics production.[27] Not to be deterred, Barshevsky correctly argued that such a law would ultimately be declared illegal, because it would violate the World Trade Organization's prohibition against quantitative restrictions on imports.[28] Even so, the legislation passed and activists cheered, knowing that the next battle might require them to confront the World Trade Organization (WTO), an institution notorious for its anti-environmental practices.

And as the high-tech industry goes global, social movement organizing against high tech must also do so. SVTC and other groups around the globe have coalesced and worked on various campaigns in Scotland, Taiwan, Ireland, and Costa Rica: "By participating in national and international forums with industry, government, academics, and grassroots activists, we are able to directly influence the debate regarding acceptable terms of the global expansion of the high-tech industry. In short, we think locally and act globally."[29]

Beyond Protest, Toward Negotiation

Environmentalists have often bypassed governments and negotiated directly with corporations to implement policy changes. For example, in 1989, a coalition of local, regional, and national environmental organizations pressured IBM into developing an aqueous-based solution as a substitute for chlorofluorocarbons (CFCs) in their operations.[30] More recently, also in response to social movement pressures, several semiconductor firms pioneered the use of natural soaps and citrus solutions in an effort to create less pollution inside and outside the factories.[31]

SVTC has collaborated with the Southwest Network for Environmental and Economic Justice (SNEEJ, an EJ organization in Albuquerque) to challenge Intel's toxic fab plant in that city. SNEEJ placed Intel's polluting practices, its resource depletion (specifically water use and pollution), its toxic workplace conditions, and the millions of dollars it received in state subsidies on the public agenda in New Mexico.[32]

SVTC and SNEEJ also joined forces to form the Electronics Industry Good Neighbor Campaign to negotiate with SEMATECH. SEMATECH is a nonprofit, taxpayer-funded consortium of U.S.-based electronics firms that Congress has supported for the purposes of maintaining the U.S. high-tech industry's competitiveness vis-à-vis European and Asian firms. SEMATECH is a coalition of eleven semiconductor manufacturers and is based in Austin, Texas. In 1992, the EIGNC persuaded Congress to earmark 10 percent of SEMATECH's $100 million taxpayer-subsidized budget for research and development of environmentally sustainable manufacturing processes in the industry.[33] At the urging of EJ groups, Congress also mandated that SEMATECH consult with environmental and labor organizations to evaluate and determine spending priorities: "By insisting that public money spent to assist private companies in developing new technologies must have a public, environmental, and labor component, the Campaign for Responsible Technology successfully changed SEMATECH's mission."[34]

While these developments are certainly reason to be hopeful, not all social movement campaigns to reform the high-tech industry's practices and policies work out so well. The Common Sense Initiative's Project XL is a case in point. In the early 1990s, under administrator Carol Browner, the USEPA developed a multi-stakeholder project for meeting the regulatory desires of industry and the environmental demands of activists. The Common Sense Initiative (CSI) was a federally supervised project that focused on six major industries, one of which was electronics. The goal, according to the USEPA, was to achieve "cleaner, cheaper, and smarter" manufacturing in each of these sectors, while simultaneously loosening up regulations that were believed to be smothering industry. In other words, CSI offered regulatory "flexibility"[35] in return for "(a) superior environmental performance; (b) transparency; (c) worker safety and environmental justice criteria; (d) increased community involvement; and (e) enforceability."[36] The stakeholders involved at various times in these projects were unions, environmental justice and community organizations, industry, and government. SCCOSH and SVTC representatives were among this group.

Project XL (or Excellence in Leadership) was the flagship initiative to emerge from CSI's electronics sector. The initiative would be located at Intel's plant in Chandler, Arizona. Intel's fab plant in Arizona uses around one billion gallons of water annually and produces significant air and water pollution. However, Intel had argued that, if the EPA were to loosen up many of its regulatory requirements, the corporation would be free to "innovate" and produce its microchips much more efficiently and with less pollution.[37] From the start, environmental and community organizations were dismayed at Intel's reluctance to disclose information about its production processes, its environmental record, its toxic materials use, and its worker safety protocols. In the name of "protecting proprietary interests and trade secrets," Intel bulldozed along while ignoring CSI's basic principles of "stakeholder participation," which involve democratic and consensus-based decision making. David Matusow, an environmentalist who participated in more than 70 of the 100 meetings around this project, recalled that Intel wielded inordinate power over other stakeholders. According to Matusow, Intel decided that "most meetings would not be open . . . that the process would not be done as a consensus, but instead provide for a voting system . . . [and] the company would not accept a local stakeholder being able to veto the agreement or an element of the agreement."[38] In this way, Intel violated the very core of the consensus-based decision-making model upon which both Project XL and the Common Sense Initiative were based.

EJ groups had also repeatedly asked that a comprehensive health study of Intel's workers be conducted, but the company made it clear that they would never allow such a study. Before long, activists realized that the project would not work and the CSI charade was over. More than a hundred community, environmental, and labor organizations condemned Project XL as a "sweetheart deal" between Intel and the USEPA that amounted to old-fashioned deregulation in disguise.[39] As further evidence that Intel was never serious about environmental reform, a short time after the Project XL debacle, in 1998, Intel refused to join the Coalition for Environmentally Responsible Economics (CERES), despite being asked to do so by its own shareholders.[40]

It is illuminating to contrast the record of voluntary initiatives like Project XL with those campaigns that involve direct-action strategies that prod, shame, and coerce industry toward change. The first strategy almost never works, and while the confrontational approach certainly cannot claim a perfect record, its success rate is much higher.

Two fundamental and irremediable flaws are embedded within most policy-driven and academic proposals for sustainability. The first is that they fail to address the core problem of power and inequality among industry, the state, workers, and communities. The second flaw is that few of these proposals are willing to challenge the capitalist growth imperative. Without addressing both of these issues simultaneously, none of these policy prescriptions has a prayer of moving us beyond symbolic reform. The best hope for this type of piecemeal reform is that it will lead to examples like IBM and HP recycling computers, or Commonwealth Edison introducing and controlling solar power,[41] or Dole Fruit Corporation producing "organic" bananas.[42] While these measures of reform are perhaps an improvement, they are only possible because they do not challenge the corporation's core values or its control over the marketplace, and so do not significantly alter the power equation between the corporation and the public. In other words, we now have recycled computers, solar power, and organic produce available from HP, Com Ed, and Dole, but only on their terms and conditions. Furthermore, these corporations do not disclose production "secrets," so we have little information about how their products are made and under what conditions workers are laboring, and consumers are at the mercy of the firm's decisions regarding pricing and quality control. This is not environmental justice or sustainable community development; it is corporate environmentalism and "green-wash" public relations—something the majority of the U.S. environmental movement's establishment is perfectly willing to accommodate.[43]

Ultimately, what EJ activists must do is redefine the terms of the debate. SCCOSH activist Raquel Sancho begins in the workplace. She describes what SCCOSH calls the "Worker Standard of Justice":

> HealthWatch [a project of SCCOSH] created a very difficult goal—to have a humane working condition that is representative of the workers' standard of justice. When you say workers' standard, that means I have an itchy nose at work because I can smell gas. But if you complain to OSHA, they will tell you that the workplace is safe up to 200 parts per million. But you say "the company was only using 190 parts per million. And look at my dress, it has holes. Does that tell you?" The workers' standard of justice means, "If I say so, I say so." The workers' standard defines how we look in the workplace, how we eat, how we breathe, how we live.[44]

Also in the tradition of redefining the terms of the debate, in 1996, the Campaign for Responsible Technology (a project of SVTC) drafted the Silicon Principles, a set of guidelines they believe the high-tech industry should endorse.[45] These include:

- Establish a comprehensive toxics use reduction program.
- Develop health and safety education programs and monitoring for the workforce.
- Work with local communities to establish "Good Neighbor Agreements."[46]

- Implement a Worker Improvement Program and Economic Impact Statements.
- Support national Research and Development policy directed by civilian (not military) needs.
- Establish corporate policies requiring equal standards for sub-contractors and suppliers.
- Establish corporate standards that are enforced equally, domestically, and internationally.
- Establish a life-cycle approach to all manufacturing, from R&D to final disposal.
- Work closely with local communities and workers to ensure full oversight and participation.[47]

While the Silicon Principles are only principles, they redefine Silicon Valley's problems and solutions and reveal an explicit challenge to the hegemony the industry currently exercises over workers, communities, and environmental policy locally, nationally, and transnationally. It is clear that, without the active presence of social movements outside and/or inside the plant gates, government and industry practices will remain unchallenged and unchanged.

Labor Campaigns

Toxics and Work

Antitoxics and pollution control measures are only one form of industry restructuring needed in the Valley. Labor protections and worker empowerment must also be at the forefront of a future sustainable and environmentally just Silicon Valley.

The links between environmental and labor advocacy are numerous. Many pro-labor and occupational health campaigns in the Valley began with environmental and anti-toxics efforts in communities. For example, in 1986, SVTC, SCCOSH, and other environmental justice groups were successful at urging the state of California to pass Proposition 65, an antitoxics initiative that forbade companies from dumping chemicals known to cause cancer or birth defects in drinking water if the chemicals exceeded safe levels.[48] The Proposition also required that individuals be warned about such chemicals when they were found at unsafe levels in the workplace or in food and other consumer items. California Governor George Deukmejian, a public enemy of environmental and labor interests, refused to enforce Proposition 65 in the workplace. In 1990, environmental justice groups sued the governor and won their case in a court of law. The measure holds employers to a standard higher than that required by federal law and was viewed as giving workers and unions more clout in getting the workplace cleaned up.[49]

The links between community-level pollution and workplace toxins were well known (see chapters 5 and 6). And although solutions to these problems are less common, there have been many notable advances. For example, many activists are making significant efforts to create jobs while reducing pollution through materials recycling and reuse. In 2000, the Silicon Valley Toxics Coalition developed a project that would pay a living wage to under- and unemployed people of color and immigrants in Silicon Valley to refurbish and repair old computer equipment for reuse.[50] The project has yet to attract major funding, but in principle, it holds great promise. This is one of many similar programs involving the repair, sale, and reuse of computers and other goods around the nation.[51] If these unemployed workers *do* get such jobs, they still need a say in how the companies operate. This question often leads to a consideration of worker advocacy groups and unions.

Unions in Silicon Valley

Advocating on behalf of electronics workers has, in many ways, proven more difficult than advocating on behalf of the environment. This is chiefly because the U.S. electronics industry excels like no other industry in the art of union-busting.[52] For example, between 1971 and 1982, out of 1,200 high-tech organizing drives in the United States, only 21 were successful, and none was in Silicon Valley.[53] Despite the superlative union-busting ability of Silicon Valley companies, unions must share a great deal of the blame for not gaining a foothold in the area. After World War II, "business unionism" took hold in the United States and was characterized by an unwritten agreement between most large labor unions and firms to exchange modest workplace reforms for peace on the shop floor.[54] Unfortunately, business rarely made good on these promises and wages and working conditions such as health care and collective bargaining rights have worsened rapidly ever since. On the whole, labor unions have not set about the task of organizing new workers, choosing instead to spend what little energy they have left among those workers already in unions. For the most part, organized labor's stance on Silicon Valley is no different.

In many ways, SCCOSH set a major precedent in Silicon Valley by organizing workers without the union label and baggage. SCCOSH provides workers with education, empowerment, negotiation skills, a "hall" for camaraderie and information exchange, a legal clinic, and an advocacy center. Other Silicon Valley groups have begun to do the same. We discussed Asian Immigrant Women's Advocates (AIWA) in chapters 5 and 6, and other groups have appeared on the terrain more recently. Silicon Valley Debug is a worker center devoted to advocating for young Silicon Valley workers, particularly temporary employees. To the AFL-CIO's credit, their South Bay Labor Council began a group called Working Partnerships in 1995. This is a union-affiliated group that has set up a nonprofit temp firm offering health insurance for temps in the Valley. Workers who join are not unionized, but they can enjoy benefits associated with unions—benefits that firms increasingly do *not* offer.[55] This strategy might be a prelude to organizing temps into unions, which has been done by the Service Employees International Union (SEIU) in San Jose via the Justice for Janitors Campaign.

Justice for Janitors

The one place unions have made significant headway in the Valley is *outside* the main sphere of high-tech production. In 1989, SEIU Local 1877 spearheaded what would become the most innovative and successful union organizing drive in Silicon Valley history: the Justice for Janitors Campaign. Like the efforts in the garment industry, SEIU took the position that the client companies (i.e., the companies in which the custodial work is actually being done)—not just the janitorial firms (i.e., the employers)—must be held responsible for working conditions. Santa Clara County has one of the largest janitorial industries in California, with an estimated 11,500 janitors in 1990. This sector ranks in the top ten in the county in terms of absolute growth. One study reports that 80 to 90 percent of Silicon Valley janitors are immigrants, mostly from Mexico and Central America.[56]

The campaign relied on a three-pronged strategy: (a) to reach out to immigrant janitors for union recruiting, (b) to target the companies where services were being performed, and (c) to use appropriate and available legal channels to push for reforms. The campaign soon found a major target—Apple Computer in San Jose. Apple secured janitorial services for its buildings by contracting with Shine Building Maintenance, a janitorial firm that was paying its workers below a living wage. For months, activists demanded that Apple negotiate with the union to allow the workers to join. There were public hearings on "the workplace abuse of immigrant janitors," a candlelight vigil at the Apple CEO's home, a phone-in campaign to the CEO's office, and a hunger strike. Finally, Apple agreed to pressure Shine Building Maintenance to negotiate with the union and allow its workers to join the union if they so desired.[57] After SEIU organized the 140 workers at Apple, they went on to successfully unionize the janitors at Hewlett-Packard in Silicon Valley. The workers received substantial pay raises and family medical benefits. John Barton, an SEIU organizer, explained that the campaign was successful not only because they had local, state, regional, national, and international support from groups pressuring Apple to negotiate a contract, but also because they emphasized the contrasts between "the immigrant female janitor versus CEOs like John Scully and the mighty Apple Computer, a David versus Goliath fight. We tried to exploit that difference at actions and events."[58]

Versatronex: Making History

While the Justice for Janitors Campaign was a major milestone in Silicon Valley labor relations, the industry could still boast that no group of workers directly involved in high-tech labor was unionized. Then, in 1992, 85 Latino workers at Versatronex, a PC board firm in Sunnyvale, did what no group of electronics workers had ever done in Silicon Valley's fifty-year history: they called a strike and walked off the job. Workers had endured racist slurs from management, inordinately low wages, no health insurance, and toxic conditions that caused nosebleeds and a host of health problems. The last straw was the firing of Joselito Munoz, an assertive worker at the plant who had spoken about these injustices at a meeting called to air shop-floor grievances.[59] At the time of the walkout, strike leaders called the United Electrical, Radio and Machine Workers Union (UE), the only union with a history of serious organizing efforts in the Valley. And, although no high-tech organizing drive had been successful in recent history, Versatronex workers knew that if they were to join a union, this would have to be the one.

Versatronex workers met with representatives of Digital Microwave Corporation (DMC), a firm that provided at least 10 percent of Versatronex's annual revenue. Workers asked DMC to stop doing business with Versatronex. If DMC refused, workers said, they would stage a hunger strike on the grounds of DMC's headquarters.[60]

Workers reported that conditions at Versatronex were characterized by a lack of ventilation in toxic spaces and that they were denied gloves and other protective equipment when they cleaned printed circuit boards with chemical solvents. Countering these claims, Versatronex's president proclaimed that "the reason for the 'walkout' had nothing to do with unsafe working conditions." In fact, he said, Versatronex "prides itself on maintaining a safe and clean working environment."[61] Instead of negotiating with the workers, Versatronex hired the nation's largest management-labor law firm—Littler, Mendelson, Fastiff, and Tichy. Striking workers argued that the company was spending more money to fight them than it would cost to reach an agreement. The law firm confirmed that it was billing Versatronex up to $290 per hour—more money in a single hour than many Versatronex workers took home after an entire week of work. But ultimately, money was not the issue: "We want respect," stated Maria Martinez, a senior assembler at the company.[62]

The walkout lasted six weeks and was brought to a halt only when the National Labor Relations Board ordered Versatronex to recognize the union and take back the seven strike leaders the company had fired. For several hours that day in January 1993, Silicon Valley appeared to be undergoing a major transformation of labor-management relations. Versatronex's unionization would set a precedent that labor would celebrate and high-tech firms would curse. However, before the end of the day, all euphoria on the part of labor advocates and unions was doused as they were reminded why Silicon Valley is union-free: Versatronex declared bankruptcy and announced that it would shut down the plant the very next month.

This incredible turn of events served a number of purposes. First, it sent a message to the investment community—the shareholders, the "market"—that Silicon Valley would continue its tradition of maintaining a "good business climate." Second, it reassured other electronics firms that no pro-union precedent would be set. The existence of just one unionized shop would have ushered in a war on labor and an accelerated shifting of production off-shore. Third, a strong message was sent to labor and EJ activists: no unions are allowed in Silicon Valley, and if you do try to organize one, firms will go bankrupt and move away rather than allow it.

The tradition of union-busting in Silicon Valley's electronics industry has deep roots. Since the late 1950s, when perhaps six unions were active in the Valley, the trade associations and individual firms have made it one of their primary missions to destroy all vestiges of worker autonomy, the most visible manifestation of which is the union. Bob Noyce, one of Intel Corporation's founders, told the author of the book *Silicon Valley Fever*, "If we had the work rules that unionized companies have, we'd all go out of business. This is a very high priority for management here. We have to retain the flexibility in operating our companies."[63]

But one labor union does have a persistent history of organizing in Silicon Valley: the United Electrical Workers Union (UE). Michael Eisenscher is a long time UE activist who was sent into Silicon Valley in 1980 to organize workers there. He shared with us the historical and theoretical basis of the UE's organizing campaigns in Silicon Valley:

> At the time there was a rank-and-file committee chaired by David Bacon at National Semiconductor that had been lobbying the union to put full-time staff in the Valley. When I arrived here, Pat Lamborn had already left National Semiconductor to launch the Electronics Committee on Safety and Health with Mandy Hawes (ECOSH, the organizational predecessor to SCCOSH). We had a committee that was multi-employer. We had a theory about organizing the Valley that it couldn't be done one company at a time. Our focus was primarily semiconductors, though we did include others. I was working with workers in six companies, which together employed upwards of 75,000 people. Our effort was to build organizing committees in each of these companies. Given the fact that this was an industry that had expended extraordinary amounts of resources to keep unions out—had been founded on the basis of no unions—we recognized that this was not going to be a quick struggle. United Electrical Workers has a long tradition going back to its founding. The whole basis on which the union is predicated is that the members run the union and that the most important point of struggle is the point of production. The further away the issue gets from the shop floor, the harder it is to settle and the less workers are involved in the settlement. The Electronics Organizing Committee was founded on that theory. It goes back to the early CIO. When the CIO organized originally, they weren't paid staff organizers going around handing out cards. Workers organized themselves. They confronted the boss and then they went to the unions and said, "By the way, we're organized. Take us in."

> We felt that that's the way you build a union. In this industry in particular, if you're going to have something that's going to survive over the long haul in a campaign that can take years, the only way you can do that is if workers become directly involved. The whole theory of organization is based on building power. The UE has been around since 1936.[64]

Eisenscher emphasized the coalition-building that occurred between the union and EJ groups in Silicon Valley during the 1980s: "There was no question that health and safety was a huge issue in the industry. The UE was supportive of the formation of ECOSH. With ECOSH taking the lead and UE supporting the whole effort to expose the fact that toxics are being used, it was a significant social issue."[65]

Eisenscher also deftly challenged the claim that Silicon Valley's multiethnic, multiracial workforce is unique and therefore somehow inappropriate for unionization (as many labor leaders have argued). And he offered a critique of the U.S. labor movement:

> The argument that all of the different ethnic and language groups in Silicon Valley are a hindrance to organizing holds absolutely no weight, if you look at history. It's no more a liability than the steel industry in the 1930s. They put out flyers in twelve languages. It can be done. I don't know any place I've ever organized that there weren't obstacles. Are you only going to organize places that are all white men? Those become rationalizations. In reality, they can be seen as assets. It depends on how you approach it. It takes more work. You've got to be sensitive to language differences, work hard in overcoming racial prejudices, and have good translators. But you can't build a movement unless you do that anyway. Any strategy for organizing in industry requires an international component, something the American movement is not very good at. It's going to take a long-term, multi-year effort, millions and millions of dollars. Most labor leaders in this country have very short timelines. They have to justify their efforts. In the 1940s and 1950s the labor movement stopped organizing. It was only in the last ten or fifteen years that it tried to re-establish its capacity to do that. It organizes for immediate results. It is tied up to the bureaucracy. There is racism. It's not just the electronics industry. We're dealing with a labor movement still largely run by white men. There is culture, racial, and gender bias. That's all begun to change, but those are still impediments. There are organizational and jurisdictional issues. In the absence of a labor investment in organizing,

> environmental and health and safety organizing became the dominant struggle in this industry. The labor movement is going to have to develop connections with environmental groups.[66]

Echoing Eisenscher's prescriptions, the AFL-CIO has proclaimed its intention to begin recruiting and organizing California immigrants—both documented and undocumented—into its unions in the twenty-first century. Given that California's workforce is mostly Asian and Latino, this makes good sense. One organizer for the Labor Immigrant Organizing Network (LION), an AFL-CIO–affiliated organization in the Bay Area, stated, "we basically feel immigration laws should be broken. We should protect undocumented workers, we should harbor them, we should not cooperate with the INS."[67] Another union organizer stated, "If everybody who was undocumented stopped working for one day, the California economy would collapse. Who do you think picks lettuce in this country? Or works in hotels, restaurants, car washes—everything we like to have at a low price?"[68] It seems that unions in California are finally realizing what Michael Eisenscher has been arguing since 1980.

But progressive change will also require the presence of willing collaborators in the state apparatus. And they are out there. One such representative of Cal-OSHA said the following to a group of immigrant workers and their families concerned about health and safety in the Valley's electronics firms:

> We only have 200 inspectors for the entire state of California. We need at least 1,000 and could use 200 in the Bay Area alone. The only reason why we have government agencies to monitor the workplace is because of the worker's movement. It's my job to enforce the regulations. But it's your job to make a new system that works for you. You have to envision it.[69]

Conclusion

Labor and environmental justice organizations have made their mark on Silicon Valley's political and economic terrain. Laws, government, and industrial policies and practices governing labor and the environment have been introduced, implemented, and changed—all largely due to the presence of active and persistent social movements. Whether they can continue to reform an industry that is growing more rapidly around the globe than any other remains to be seen.

10

The Broader Picture

Natural Resources, Globalization, and Increasing Inequality

The Importance of Natural Resources

Many scholars and pundits have proclaimed this era of high technology the "Information Age," characterized by the "death of distance," "virtual economies," "dematerialization," and "weightlessness."[1] Each of these claims conjures up images of high-tech society liberated from the constraints of nature—space, time, gravity—that have defined the boundaries of all previous civilizations. We must make an observation that may seem obvious to many of us but is increasingly less so for others: in this so-called Information Age or Virtual Age, natural resources—particularly land, minerals, fuels, and water—are necessary factors for our survival and are the root input factors in the global economy. As one author puts it, *Materials Matter*, because as abstract and virtual as many of us would like to believe the Computer Age may be, all of our machines and technologies are made using petroleum, water, metals, and chemicals that are earth-based and finite.[2] We are as natural resource–dependent as human society has ever been. In fact, that dependence is deepening as we increase our use of ecosystems to fuel the global economy. For example, between 1992 and 1998, U.S. imports of finished steel jumped from 13 million to 35 million net tons,[3] and between 1985 and 1996, Latin American nations extracted and exported 2.7 billion tons of basic resources, most of them nonrenewable.[4] The United Kingdom exhibits a similar pattern:

> In the UK, iron consumption has increased twenty-fold since 1900; the global production of aluminum has risen from 1.5 million tonnes in 1950 to 20 million tonnes today [1999]. In the decade 1984–1995 (during a period in which we should have seen the "weightless" effect

> become visible, if the theorists are to be believed) aluminum consumption in the UK rose from 497,000 tonnes to 636,000; steel consumption increased from 14,330,000 to 15,090,000 and wood and paper consumption more than doubled, from 41 million to 93 million tonnes.[5]

Echoing this sentiment and applying it to Silicon Valley's industries, activist Jay Mendoza writes:

> The Information Super-Highway uses huge amounts of energy to power millions of network servers and all the computers and computer peripherals associated with it. To power the Internet, enormous amounts of electricity are generated from oil and gas-fired power plants, which spew tons of toxic pollutants into the air, poisoning those who live near them. The biggest vampires of energy are huge data server farms that house hundreds of computers on racks inside several stories of a building [one is currently being proposed for the working-class Latino community of Alviso]. Investigative research has uncovered a link between oil energy providers, such as Enron Corporation [previously the largest energy conglomerate in the United States], who are hooking up with high-tech companies such as IBM and AOL, to make electricity deals and sell oil-based energy over the Internet to AOL users. The ignored cost is the human health impacts of building power plants in neighborhoods populated by low-income minority peoples.[6]

Finally, the extraction of one natural resource in particular—coltan—which is vital to the construction and operation of many electronics components, is threatening ecosystems and indigenous peoples throughout eastern Congo. More than twenty illegal coltan mines are in

> the Okapi Faunal Reserve, a protected area in the Ituri rain forest. . . . The reserve is named after a reclusive, big-eared relative of the giraffe that is found only in Congo. Along with about 4,000 okapi, the reserve is home to a rich assemblage of monkeys (13 species), an estimated 10,000 forest elephants and about the same number of Mbuti people, often called pygmies, who live by hunting, gathering and trading. . . . Coltan is abundant and relatively easy to find in eastern Congo. All a miner has to do is chop down great swaths of the forest, gouge S.U.V.-size holes in streambeds with pick and shovel. . . . Coltan is the muck-caked counterpoint to the brainier-than-thou, environmentally friendly

> image of the high-tech economy. The wireless world would grind to a halt without it. Coltan, once it is refined in American and European factories, becomes tantalum, a metallic element that is a superb conductor of electricity, highly resistant to heat. Tantalum powder is a vital ingredient in the manufacture of capacitors, the electronic components that control the flow of current inside miniature circuit boards. Capacitors made of tantalum can be found inside almost every laptop, pager, personal digital assistant and cell phone.[7]

Nothing about the high-tech global economy is "weightless" or "virtual."

As the global economy marches into the twenty-first century, depleting and contaminating more and more natural resources without pursuing alternative sources of energy and material inputs, we have become more susceptible to economic and ecological catastrophes than ever before. Because of this heightened vulnerability and intensified dependence on natural resources, some scholars have argued that activists who oppose those corporations directly involved in the extraction of oil and minerals, for example, are perhaps much more threatening than one might think. This is because when activists challenge these extractive corporate activities, they are threatening the very lifeblood of the global economy.[8]

Short of a major challenge to corporate power, less radical proposals for reform abound. For instance, many activists and scholars have demonstrated that, given the necessary political will and means, even the most toxic and polluting industries could substantially reduce their chemical inputs and outputs. If manufacturers, suppliers, governments, workers, and customers were all to agree on more socially and environmentally responsible policies for production, these goals could be met and sustained.[9]

The recent and historical successes and setbacks of the environmental justice movement in Silicon Valley must be placed against a backdrop of rising political, economic, and social discord around the planet. At root is the growing disparity between the wealthy and the poor and between corporate power and indigenous peoples, people of color, and the working class within and between nations. Social inequality in the United States and around the globe has increased dramatically in the past three decades.[10] This growing gap between social groups is largely a result of the acceleration of transnational economic activity, which has pitted individuals, social classes, racial and ethnic populations, neighborhoods, cities, regions, and even nation-states against each other in a competitive world system.[11] Formal commitments to international trade accords like the North American Free Trade Agreement (NAFTA), the Free Trade Area of the Americas (FTAA), and the World Trade Organization (WTO) have created great uncertainty for those social movements hoping to build environmentally just and sustainable communities. The goal of all free trade agreements is to ensure the free and unencumbered flow of capital and currency without regard for laws, culture, tariffs, taxes, and the like. In short, free trade agreements and their proponents are openly antagonistic to concerns about labor, community development, and ecology. And while EJ activists may be tempted to continue targeting city councils, state legislatures, and the U.S. Congress, all of these governmental bodies are now ultimately serving at the pleasure of the WTO. In other words, with the United States's entry into the WTO, the laws, ordinances, and policies of the 39,000 different governments (states, counties, cities) within the United States and the other member nations of the WTO must now conform to the WTO's rulings. So while the functioning of local, state, and national governments is crucial for daily governance, we now have additional institutions to confront, whose core principles and practices devalue people and the ecosystem, spawning environmental injustices with virtually every ruling. Our hope is that the environmental justice movement in Silicon Valley will continue to work with labor, human rights, women's rights, and global justice movements to directly target the WTO and all free trade agreements, and the transnational corporations controlling them. We also caution activists that any movement that places its focus solely on governments, the courts, and the traditional political process will fail. So when we think about the many laws, ordinances, and policies that activists in Silicon Valley have pushed for, we must also realize that each of these victories may one day be overturned for violating World Trade Organization rules.[12] Environmental Justice and sustainable community development in Santa Clara County will never be achieved without prioritizing a value system that honors people and the environment before profits.

Understanding Environmental Injustice

Immigrants and people of color have confronted environmental racism in the region now called Silicon Valley for more than two hundred thirty years in a number of ways: (a) they have frequently had their land and natural resources taken from them and destroyed; (b) they have often been denied citizenship and therefore formal political power; and (c) they have been concentrated in enslaved, indentured, and related "free" exploitative labor markets where wages are nonexistent or very low and the risks to one's health are substantial. Because human and natural resources were so heavily degraded and exploited during each phase of Santa Clara County's history, it is clear that these two factors are of central importance to the county's future survival. It follows from this that the sustainability of the Valley will depend on the empowerment of women, people of color, and immigrant workers and on respect for the area's ecosystems. Moreover, when we argue for "ecosystem" or "environmental" protection, we must not constrain our scope to the general natural environment; rather, we must extend it specifically to include the environments where immigrants, women, and people of color live, work, and play. For these are the environments that have been most targeted for theft and despoliation. From the Ohlone native people, to the Gold Rush and mercury miners, to the cannery and orchard workers, to the high-tech laborers, women, immigrants, and people of color in the Valley have always played the role of the "canary in the coal mine."[13] The rest of us should therefore view these groups as populations who have earned the right to equal—if not extra—protection, for this can only benefit the population as a whole.

Throughout this book we have used these historical and contemporary struggles to illustrate four theoretical arguments. First, we demonstrated that environmental injustices impact immigrant populations as much as, if not more than, they do communities of color. In fact, the economic and political vulnerabilities that working-class and undocumented immigrants face are compounded by their environmental vulnerabilities in workplaces and communities. One significant aspect of this finding is that a much broader base of people suffers from environmental inequalities than was previously thought.[14] And while many scholars of the "ethnicity school" might want to believe that these environmental injustices will slowly disappear as these immigrants assimilate into American culture, this is not likely to occur since the majority of immigrants today are people of color.[15] The link between immigration and the environment is perhaps most important because of the longstanding campaign by many whites to attribute a host of ecological ills to immigrant communities. Thus the finding that immigrants actually *suffer* from—

rather than contribute to—environmental degradation should challenge these nativist assumptions.

Second, we have focused on the study of the workplace as a site of struggle against environmental racism. The scientific evidence reveals that the workplace is often the source of toxics and pollution that eventually reach residential communities. As many activists and workers we featured in chapters 5, 6, and 7 made clear, if society is serious about preventing community-level environmental exposures, then we must begin in the workplace. As with the immigrant-environment nexus, the recognition of workers as the first line of toxic exposure broadens the population of people normally considered among those suffering environmental injustices. This is also an enormous constituency that the EJ movement could enlist in various campaigns.

Third, we examined the role of gender in environmental justice conflicts and found that, whether the focus is men or women, gender matters a great deal in workplace exposure to toxics and other hazards and in community mobilizations against environmental injustices. Gender has long been neglected in the EJ literature as a factor in environmental exposure, for methodological reasons. This is where each theoretical contribution we are making here is linked: when one studies working-class immigrant communities, one is naturally led to the workplace; and when one studies low-wage immigrant workplaces in the United States, one cannot help but notice that women make up the majority of employees in many of these exploitive sectors. The presence of toxics and other environmentally harmful conditions in female-dominated occupations is as commonplace in Silicon Valley's high-tech sweatshops as it is in New York's garment sweatshops. Gender and environmental inequalities are inextricably bound together.

Fourth, we investigated the global nature of environmental racism—specifically, the social and environmental costs of the transnational electronics industry. Our analysis of high tech's global reach indicates that patterns of worker exploitation and ecological disruption we observe in Silicon Valley have been replicated and made worse in Latin America, the Caribbean, parts of Europe, and Asia. The significance of this finding is that the fate and health of communities and workers in the United States are tied to communities and workers around the globe. Environmental justice activists and scholars can no longer view pollution and social exploitation in the United States independent of its origins and/or impacts on people and ecosystems elsewhere.

The movement for Environmental Justice and scholarly studies of EJ struggles must place a greater emphasis on immigrants, the workplace, gender, and globalization. Indeed, probably all EJ conflicts occurring in the world touch on most of these factors. Whether it be sweatshop labor, oil and mining struggles, toxic waste dumping, transportation racism, poor housing quality, lead poisoning, indigenous land and sovereignty conflicts, or water rights disputes, all of these are likely to involve many of these issues. Unfortunately, we as activists and scholars have been slow to recognize the importance of each of these factors individually or at their points of intersection. We believe this is largely because traditional models of environmental justice conflicts were grounded in local toxic facility siting disputes in native-born residential communities, where activists drew on resources, language, and tactics inspired by the civil rights movement.[16] This focus largely ignored immigrant communities, workers, the role of gender as a factor in exposure to hazards, and extralocal (i.e., transnational) forces that may have shaped the nature of the environmental harm in question. While there have always been exceptions to this rule, facility siting remains the principal emphasis of the movement and the scholarship around it.[17] Placing the four factors we have considered above closer to the center of the analysis is crucial to building a stronger movement for environmental justice and to understanding the full context and contours of environmental inequalities.

Coming Full Circle

In our discussion of the Ohlone culture and civilization, and, ultimately, their genocide, the reader may have been left with the impression that all members of this Native American nation are now extinct. This is not the case. Throughout the twentieth century, descendants of the original Ohlone inhabitants of Santa Clara Valley (and nearby areas) have made their presence known. As with environmental and labor struggles in the Valley's high-tech industry, Native Americans have demonstrated that social movement advocacy is the best way to achieve results. In 1928, Ohlone folk were involved in a legal suit brought against the U.S. government by California Native peoples seeking reparations for stolen lands. Ohlones and members of other Native nations filed another such lawsuit in 1964. During the 1960s, the San Francisco-based American Indian Historical Society began researching, publishing, and advocating for the Ohlone people. They were successful at preventing the destruction of a mission cemetery (in the path of a proposed freeway) containing the remains of many Ohlone people. Descendants of the Ohlone involved in this case incorporated as the Ohlone Indian Tribe, and they now hold title to the cemetery in Fremont, California.[18] In 1975, a related incident occurred south of Santa Clara County. Workers discovered an Ohlone burial ground during the construction of a warehouse in Watsonville. Local Native people took up arms, barricaded themselves at the site, and eventually halted construction. The Pajaro Valley Ohlone Indian Council was formed to oversee the burial ground's future protection.[19]

New organizations, such as *Muwekma*, have emerged to advocate for self-determination and the federal recognition of Ohlone people. While federal recognition is a slow process, the Ohlones have successfully used the environmental impact assessment requirements of the National Environmental Policy Act (NEPA) and the National Historic Preservation Act in their latest struggles. For example, in the City of San Jose, just yards from the Guadalupe River, a cannery operated from 1918 into the 1960s. In 1989, on that same spot, a light rail transportation terminal was being built. Early in the construction phase, builders discovered a burial ground containing more than 124 Ohlone persons. Construction was immediately halted, as required by law, and Ohlone descendants arrived to supervise the proceedings. Eventually, all the bodies were reburied deep in the ground near the original site. The light rail terminal is now called the Tamien Station, because it also stands on what was originally Tamien Ohlone land and a portion of Chief Lope Inigo's land grant, Rancho Posolmi. This example illustrates the deep relationship among Santa Clara County's first peoples, the integrity the local environment, and the future of the Valley as a multiracial society.

We would like to close with the words of a descendant of the Valley's original inhabitants. She is an Ohlone woman, Ann Marie Sayers:

> This country is built on the lives and deaths of Indians who are environmentally aware; whose societies operated within traditional systems of authority and law, and who lived culturally rich and deeply spiritual lives centuries before the Spaniards arrived. Until this is recognized and there is *truth in history*, the life of the natural world, which is the real foundation of the well being of this country, will continue to be lost.[20]

Notes

Notes to the Preface

1. Environmental *justice* is a vision of a future in which environmental racism/inequalities no longer exist and in which a given community is characterized by the prevalence of social justice and ecological sustainability.

2. Bryant 1995.

3. In our analysis we try not to create a hierarchy that privileges one form of social inequality (e.g., race, class, or gender) over another. These factors exist in relation to one another, are interactive, and are often context-specific (King 1988). Thus race does not determine class position, but identifiable patterns of social inequality exist that place certain groups in specific social spaces over time, via institutional discrimination. It is certainly the case that, independent of class and gender, people of color generally experience racism on a collective and individual basis. However, the impact of class and gender shapes the nature of that experience, its frequency and intensity. Furthermore, the contributing factors that shape the experience of racism, classism, and sexism/patriarchy change over time.

Notes to Chapter 1

1. Arnold 1976, 26; Findlay 1992, 143.

2. Generally, when we refer to "Silicon Valley," we are referring to an area that roughly corresponds to Santa Clara County, the Santa Clara Valley, and parts of San Mateo County. Silicon Valley firms, and electronics/semiconductor firms in general, are not restricted to this area. Many are now located north in Oakland, in San Francisco, to the east in Dublin and Pleasanton, and further south into Santa Cruz and San Mateo County. Furthermore, Silicon Valley workers often commute to Santa Clara County from three other surrounding counties—San Mateo, Alameda, and Contra Costa. Santa Clara County's population is now 1.7 million, and the Valley's population is 2.3 million (see Silicon Valley Environmental Partnership 1999). Only industry insiders used the term "Silicon Valley" until around 1971, when software engineer Don Hoefler popularized it in a trade journal (Johnston 1982).

3. Malone 1985.
4. Page 2000, 37.
5. D'Souza 2000.
6. Markoff 1999.
7. Hammonds 2000.
8. Ibid.
9. Markoff 1999.
10. Ibid.
11. Hansen 1982, xi.
12. Carey 1986.
13. Ibid.
14. Hayes 1989.
15. Glenn 1992; Zinn 1994.
16. Bryant 1995.
17. Early studies focusing on the association between social class and pollution included: Asch and Seneca 1978; Berry 1977; Freeman 1972; Schnaiberg 1973, 1980, 1994. Research that was more explicit about race and pollution included: Bryant 1995; Bryant and Mohai 1992; Bullard 2000, 1993, 1994; Cole 1992; United Church of Christ 1987; Schnaiberg and Gould 1994; Swanston 1999; Taylor 1993.
18. Chavis 1993.
19. Lavelle and Coyle 1992; United Church of Christ 1987; United States Environmental Protection Agency (USEPA) 1992a and b; United States General Accounting Office 1983. And although a small group of scholars conclude that racial inequities are far less evident than environmental inequalities along socioeconomic lines in hazardous facility siting, or that impartial market dynamics—not discrimination—are the root of environmental injustices (see Anderton et al. 1994; Been 1993; Coursey 1994), the great majority of environmental justice research confirms the environmental racism thesis.
20. See Bullard, Grigsby, and Lee 1994; Bullard and Johnson 1997; and Bullard, Johnson, and Torres 2000.
21. Faber and Krieg 2001.
22. Daniels 1997, 74, 83; Rumbaut 1996, 26. Immigration restrictions have also occurred throughout U.S. history, including hallmark years like 1882, 1924, 1952, 1986, 1994, and 1996.
23. Daniels 1997.
24. Rumbaut 1996, 29.
25. Another distinction is the explicitly class-bifurcated nature of recent immigration, particularly among Asians. One segment of this immigration flow is economically "well off," while the other segment is working-class to working-poor.
26. Fujiwara 1998.

27. Bustamante 1997; Fujiwara 1998; Sanchez 1997; Smith 1998; Wilson 1997.

28. Bustamante 1997; Chan 1986.

29. Smith 1998, 3;Wilson 1997, 51.

30. Kwong 1997, 16; Su 1999, 245.

31. See Amott and Matthaie 1991; Bustamante 1997; Rumbaut 1996, 3; Sanchez 1997; Zlolniski 1994.

32. Su 1999, 250.

33. Dittgen 1997.

34. Gottlieb 1993, 2001; Hunter 2000; Hurley 1995; Perfecto and Velasquez 1992.

35. See Friedman-Jimenez 1994; Gottlieb 2001; Hurley 1995; Perfecto and Velasquez 1992; and Pellow 2002 for exceptions.

36. For exceptions, see Adeola 2000; Marbury 1995; Mpanya 1992.

37. See Brimelow 1995; Sferios 1998.

38. For the definitive anti-immigration treatises, see Huddle 1985, 1993, and 1994.

39. Chang 2000; Dittgen 1997, 267. California's Proposition 187 was approved by voters in November 1994. This legislation would have banned public education, public social services, and nonemergency public health care to undocumented immigrants in that state. Judged unconstitutional, the legislation was never enacted (Fujiwara 1998).

40. Espenshade and Belanger 1997.

41. Ibid.

42. Dittgen 1997, 260.

43. See Amott and Matthaie 1991; Passel 1994; Passel and Clark 1996; Sanchez 1997.

44. See Almaguer 1994; Amott and Matthaie 1991; Gottlieb 1993.

45. See Gottlieb 1993.

46. Brimelow 1995, 9.

47. For astute analyses of the U.S. conservation movement's racist, classist, and anti-immigrant history, see Gottlieb 1993 and Jacoby 2001.

48. Sferios 1998.

49. Abbzug 2001; Amott and Matthaei 1991; DeFreitas 1995; Sanchez 1997, 13.

50. For studies on environmentally contaminated immigrant communities, see Diaz 2001; Gottlieb 1993; Hunter 2000. For studies on contaminated workplaces where immigrants are concentrated, see Hurley 1995; Pellow 1999a; Pellow and Park 2000; Pellow 2001.

51. Kadetsky 1993.

52. Smith and Woodward 1992, 13.

53. LaDou 1985.

54. Bullard and Wright 1993; Chavez 1993; Friedman-Jimenez 1994; Johnson and Oliver 1989; Robinson 1991.

55. Perfecto and Velasquez 1992.

56. Castleman 1987; Roberts 1993.

57. Robinson 1991.

58. Berman 1978; Kazis and Grossman 1982; Nelkin and Brown 1984; Roberts 1993.

59. Navarro 1982.

60. Rios, Poje, and Detels 1993; Robinson 1991.

61. See Brown and Ferguson 1995; Brown and Mikkelsen 1990; Krauss 1993.

62. See chapter 3; also Siegel and Borock 1982; and United States Commission on Civil Rights 1982.

63. Hossfeld 1988; Park 1992.

64. See chapters 4 and 5.

65. Gurtunya 1992; Hossfeld 1988, 1990; Katz and Kemnitzer 1983, 1984; Keller 1983; Snow 1983.

66. Baker and Woodrow 1984; Hossfeld 1991.

67. Baker and Woodrow 1984. GABRIELA is a Philippines-U.S. transnational women's activist network that has brought advocates together to combat the global electronics industry.

68. Brown and Ferguson 1994; Brown and Mikkelsen 1990; Kraus 1993.

69. The importance of environmental justice as the struggle to protect where we "live, work, and play" has always been emphasized in the EJ movement (see Alston 1990). A few scholars have made these connections clear as well (see Kazis and Grossman 1982; Roberts 1993, 1996; Sheehan and Wedeen 1992).

70. Cf. in Rogers and Larsen 1984.

71. Levander 1992. The percentage of women on the Valley's corporate boards of directors is one of the lowest of any region in the United States.

72. Chang 2000; Roberts 1995. Many immigrants are transnational in one sense before they migrate to the United States because of their intense exposure to U.S. culture via television and consumer products (see Foner 1998). Others are transnational as a result of their experiences with travel, business, and even politics with dual citizenship and dual-country voting.

73. *Wall Street Journal*, August 28, 1998.

74. Min 1996, 303.

75. The U.S. military is the single largest producer of toxic waste in the nation.

76. Rumbaut 1996, 318.

77. More generally, adverse ecological conditions, including natural disasters, have also served to drive human migrations across nation-state borders. For example, many Chinese immigrants came to the United States during the nineteenth century as a direct result of floods that wiped out entire local economies (Amott

and Matthaie 1991). Similarly, soil deterioration in certain agricultural communities in Mexico forced many farmers northward to *la frontera*, the U.S./Mexico border, during the twentieth century (Wilson 1997, 48).

78. See chapter 8.

79. Kenney 2000; Lee et al. 2000; Matthews and Cho 2000.

80. Anthony 1928; Matthews 1976; Zavella 1987.

81. Gurtunya 1992; Hossfeld 1988; Keller 1981; Park 1992. Karen Hossfeld's *Small, Foreign, and Female* (forthcoming) is a seminal study of female immigrant labor exploitation in Silicon Valley. Our book also contributes to the growing literature on immigrant women's work in a changing economy. Some notable recent books on this subject include Chang 2000, Hondagneu-Sotelo 2001, and Lopez-Garza and Diaz 2001. This study extends our understanding of immigrant working conditions by including the environmental concerns related to a specific industry and by expanding the analysis to a variety of ethnic groups and presenting their stories in a historical, relational context.

82. Bernstein et al. 1977; Siegel and Borock 1982; Siegel and Markoff 1985; Smith and Woodward 1992. Mazurek (1999) presents an unflattering assessment of the global electronics industry's environmental record.

83. One exception to that trend is a study by two sociologists, Andrew Szasz and Michael Meuser (see Szasz and Meuser 2000), in which they analyzed social inequality indicators in Silicon Valley industries and the spatial distribution of toxic industrial emissions by race and class. The authors found that environmental inequalities by race and class were evident in the area beginning in 1990. While the authors of this important study found evidence of environmental racism emerging only in the 1990s, we demonstrate that, in fact, communities of color and immigrants in the Valley have experienced this phenomenon since the eighteenth century (see chapter 2).

84. For example, Charles de Gaulle and Queen Elizabeth both paid official visits to the Hewlett-Packard Corporation during their administrations (see Findlay 1992, 137).

85. Saxenian 1985b, 102.

86. Sachs 1999.

87. Fellmeth 1973; Sachs 1999.

88. Clune 1998, 120.

89. It is commonly agreed that, in the absence of its immigrant labor, the state of California's economy would collapse (see Fan 2000). As a measure of its intense dependence on natural resources in this "postindustrial," "postmodern," and "Digital Age," the hike in energy prices during 2001 had Silicon Valley businesses screaming for relief and threatening to close down and move offshore where regulations are more "business friendly" (see *New York Times*, January 22, 2001).

90. Working Partnerships USA 1998, 12.

91. Almaguer 1994; Chavez 1993; Kwong 1997; Pellow 2002; Smith 1998; Sferios 1998; Su 1999; Tilly 1996.

92. Amanda Hawes is one of the nation's foremost workers compensation attorneys with an emphasis on toxics and occupational health. She has represented injured and ill workers in the agriculture and high-tech sectors since the early 1970s. Hawes is also a cofounder of the Santa Clara Center for Occupational Safety and Health (SCCOSH), one of Silicon Valley's first EJ organizations.

93. Fernandez-Kelly 1983; Nyden, Figert, Shibley and Burrows 1997; Park and Pellow 1996; Stoecker 1996; Stoecker and Bonacich 1992.

94. One of the authors served on the SCCOSH Board of Directors and on the Advisory Board of the Health and Environmental Justice Project, a SCCOSH-SVTC collaboration. That same author worked as a cable assembler in a Silicon Valley firm for several weeks during the winter and spring of 1999. The other author has advised SCCOSH on issues concerning immigrant women workers and helped gather data for SCCOSH's 2001 publication of its *Workers Story Process*.

Notes to Chapter 2

1. Fox 1991, 44; Smith and Woodward 1992, 18. The Superfund program is a federally designated pool of money for the cleanup of the nation's most toxic and polluted plots of land. The program was initiated in large part as a response to the pressure brought to bear on the USEPA in the aftermath of the Love Canal disaster but has become its own bureaucratic and political disaster. The Signetics Superfund site has contaminated local groundwater with freon, trichloroethane, trichlorethylene, perchloroethylene, xylene, and butyl acetate (U.S. Department of Health and Human Services 1989b).

2. Szasz and Meuser (2000) argue that environmental racism in Silicon Valley is a more recent phenomenon, traceable only as far back as the early 1990s. We agree with their assessment, given the restricted definition of environmental racism they use (e.g., semiconductor facility siting and associated pollution). However, our study casts a much broader operational and historical net for what constitutes environmental racism and therefore builds upon Szasz and Meuser's earlier work.

3. Levy 1978. The term *Ohlone* is a contemporary correction to the many names that have been used to describe this group of Native Americans in the past. Costanoan is derived from the Spanish word *costenos*, meaning "coastal people." Some Ohlone persons living today take offense at the use of the term Costanoan because it was imposed upon their people (see Leventhal et al. 1994). And since the 1960s, many descendants of the original peoples of the area have chosen to use the term Ohlone, and journalists and scholars have followed suit (Teixeira 1997, 4).

4. Rawls and Orsi 1999, 211.

5. Cook 1951, 174.
6. Leventhal et al. 1994.
7. Cook 1951.
8. See Dasman 1999, 107. We note that scholars have documented numerous instances in the historical record where Native communities exhausted or damaged natural resources (see Krech 1999; Preston 1998). Even among the Bay Area peoples there were private holdings of small parcels of land, trees, and fishing spots—exceptions to the "communal ethic" we discuss in this chapter, which may have led to certain abuses and mismanagement of resources. However, there is simply no comparison between Native and European peoples' impacts on the North American environment. Both in ideology and in practice, the European approach to and impact on the ecosystem were unique. The magnitude, intensity, and permanence of European peoples' destruction of natural resources on the planet are unparalleled in human history.
9. Parkman 1994, 55
10. Heizer and Treganza 1970, 297, 300, emphasis added.
11. Ibid., 299.
12. Parkman 1994, 54.
13. Heizer and Treganza 1970, 302.
14. Simmons 1998, 48.
15. Preston 1998, 265.
16. Ibid., 266.
17. Crespi, in Geiger 1951.
18. Crespi, in Galvin 1971.
19. King Carlos (or Charles) III of Spain and Don Felipe Neve, the Spanish Governor of California, gave the directive to found and develop San Jose, four miles south of the Mission Santa Clara. For the next century and a half, San Jose would be viewed as a satellite or suburb of Alta California's major metropolis, San Francisco.
20. Scott 1959, 12–13.
21. Heritage Media Corporation 1996, 1.
22. Hackel 1998, 116.
23. Jackson 1994, 229.
24. Conditions were so horrible at the Mission Dolores, for example, that the governor launched an investigation into the treatment of neophytes. The report documented harsh labor conditions, inadequate food supplies for workers, and extreme cruelty on the part of the "disciplinarian" Padre Danti (Lasuen 1965, 401).
25. Webb 1952.
26. Vancouver 1801, 32.
27. Hackel 1998, 123.
28. McCarthy 1958, 139–140.

29. See chapters 3 through 7 for discussions of gender inequalities in the post-contact eras leading up to and including the high-tech epoch of Silicon Valley.

30. Castillo 1994, 282. Also see Overland Monthly 1869.

31. Castillo 1994, 283.

32. Castillo 1989; Hurtado 1992.

33. Jackson 1992.

34. Jackson 1994, 244. After visiting a *monjerio* at one of the nearby California missions and being overwhelmed by the stench of human feces, Diego de Borica, a military governor, wrote of his concern for the well-being of these Native women. It is also paramount to acknowledge the historical and contemporary racism that has served to justify the treatment of the Ohlones by (a) claiming that the Spaniards were "saving the Natives from themselves," (b) blaming the Ohlones for any negative consequences stemming from their colonization; and (c) simply excluding any mention of the Ohlone people from most of the written works about the San Francisco Bay area's history. For example, one author writes, "the brutish Costanoans arrived on the shores of the San Francisco Bay, to spend the long aeon of intervening centuries in continuous unameliorated savagery, until mercifully lifted up from their degradation by their Franciscan teachers" (McCarthy 1958, 11–12). In another passage, the same writer points out that "Fathers Duran and Fortuni of Mission San Jose list as 'dominate vices and faults' of the East Bay Indians, anger, immorality, stealing, and lying; they attribute as the principal cause of the low birth rate and crime of abortion—'common among their women'" (McCarthy 1958, 33–34). This claim that Native behavior and "vices" were the root cause of the low birth rate is of course unforgivable and no less offensive than the assumption that the Spaniards had a monopoly on honesty, morality, and generosity.

35. Between 1799 and 1835, the mean life expectancy at birth for Native persons at the San Jose Mission was "extremely low, a mere 1.7 years. Death rates were consistently higher than birth rates . . . and, as indicated by the low life expectancy, rates of infant and child mortality were extremely high" (Jackson 1994, 239).

36. Castillo 1994, 278. The San Francisco padres requested from the governor a military campaign to return the supposedly blissful, yet misguided, apostates to the mission.

37. Jackson 1994, 236.

38. Leventhal et al. 1994, 306.

39. Holterman 1970, 44.

40. Jacobson 1984, 26.

41. "Outsiders" at this time meant Anglos like Jedediah Smith, an Anglo trapper who defied the Californio authorities when they demanded that he leave California and return to the United States. Smith was friendly with many Native American nations, and scholars have speculated that he may have inspired or as-

sisted in the rebellions of 1828–1829 (see Cook 1976; Shoup and Milliken 1999, 94).

42. Keller 1981.

43. McWilliams 1949.

44. In many ways this transition parallels the shift wherein African Americans moved from slavery into the sharecropping and domestic service economy after emancipation.

45. Keller 1981, 20.

46. Dasman 1999, 106.

47. Castillo 1994, 280.

48. Leventhal et al. 1994, 306.

49. Jacobson 1984, 27.

50. Preston 1998, 281.

51. Preston 1998.

52. Keller 1981.

53. Nunis 1964, 23.

54. The parallels to the discourse around Silicon Valley's riches are striking, especially as applied to immigrants and the American Dream (see chapter 1).

55. Rohrbough 1997, 206; subsequent citations in this paragraph are by page number.

56. Jung 1999, 63.

57. Holliday 1999, 142.

58. Named after the world's largest quicksilver mine in Spain, the Almaden Mine. Of course this cave had been known to and used by Ohlones for centuries prior to Castillero's "discovery" (McWilliams 1949).

59. Limbaugh 1999, 29. For a detailed description of gold processing with a quicksilver machine see Clark 1967, 189–190.

60. See Omi and Winant (1994) for the segmented labor market theory—the hypothesis that capitalists divide and conquer workers of different race/ethnic backgrounds to keep all of their wages low and to prevent collective solidarity, despite their obvious common interests. Regarding the racial division of labor among Gold Rush workers, as J. Ross Browne wrote in the first official report on California's mineral resources, "Mexicans . . . are found to be more adventurous than Cornishmen, and [are] willing oftentimes to undertake jobs which the latter have abandoned" (McWilliams 1949, 138).

61. McWilliams 1949, 140.

62. Ibid.

63. Kennedy 2001.

64. Randol 1892, 237–240.

65. See Erikson 1994, 35–39.

66. Heritage Media Corporation 1996, 70.

67. Ibid.

68. Ibid., 78.

69. Ibid., 77.

70. See Reisner 1986 for an astute account of the state's tribulations with this precious resource.

71. Johnson 2000.

72. Mann 1982.

73. This legislation was a blatant violation of the Treaty of Guadalupe Hidalgo, which mandated that property titles held by Californios be respected unconditionally.

74. Pitt 1966.

75. This is a racial hierarchy quite distinct from the contemporary pecking order, which generally starts with whites on top, followed by certain Asian American groups, certain Latino groups or African Americans, and Native Americans (see Omi and Winant 1994; and see Park and Park 1999 for a critique of racial hierarchies and how U.S.-based sociologists have perpetuated a number of myths such as the "black-white dichotomy").

76. Almaguer 1994, 8.

77. Rohrbough 1997, 12.

78. McWilliams 1949, 137.

79. Mexico has a long history of mining (McWilliams 1949, 143).

80. Other Latin Americans, such as the many Chilean miners in the area, received equal or harsher treatment, including officially sanctioned executions (Rohrbough 1997, 224).

81. Stanley 2000.

82. Amott and Matthaei 1991, 196.

83. Almaguer 1994, 165; Amott and Matthaei 1991, 197.

84. And this was a part of the much larger conflict in the United States as a whole between slave states in the South and the "free" Northern states—a conflict that would play a role in sparking the Civil War.

85. Almaguer 1994, 12. This included African Americans, Chinese, Japanese, and Mexicans. There were in fact some free African Americans who worked the mines, but they were in short supply. Of course the irony for the white working class is that they narrowly focused on keeping paid labor out of California instead of supporting freedom for *all* workers, which could arguably have created a much stronger political force among working peoples. The whites insisted on defining and constructing the "working class" as exclusively white. While the free labor movement eschewed the opportunity for multiracial organizing, it was successful at unifying the white working class (see Almaguer 1994; Holliday 1999, 282; Roediger 1999).

86. Cornford 1999, 85; Sandmeyer 1972, 80. The historian Sucheng Chan refers to these conditions as "semi-free" (see Chan 1986). While those Chinese workers who came to the United States were not actually a part of the coolie

trade, which was centered in the Caribbean and Latin America (Chan 2001), they generally worked under contract or indentured labor conditions and were commonly referred to as "coolies."

87. After the Civil War, African Americans were eligible for citizenship via a revised Nationality Act of 1790, a law that previously restricted this privilege to "free white persons." The phrase "persons of African descent" was added to the legislation after Emancipation, allowing African Americans to join the ranks of citizens while explicitly excluding Asians. Some African Americans also joined the anti-Chinese revolt, perhaps seeking to shift the onus of discrimination away from blacks and to share in the spoils of reduced economic competition. This of course was folly, as whites continued to discriminate against blacks as they had always done.

88. Almaguer 1994, 166; Kwong 1997, 143. Chinese workers were viewed as docile and highly exploitable. They were routinely used in the most dangerous tasks in railroad building. For example, whenever the Central Pacific Railroad had to negotiate a mountain, they would dynamite their way through. In order to do this, Chinese workers were lowered into holes drilled deep within the mountain. There they would place and light a dynamite fuse and tug on the rope in the hopes that their coworkers would pull them up and out in time to run clear of the explosion. Not surprisingly, hundreds of men died in these operations. Other Chinese railroad workers were buried in mountain avalanches during the winter months and were not found until after the spring thaws. Mining for gold and other minerals continued long after the 1850s, and Chinese workers who had survived the railroad work were often employed in the most "toilsome, dangerous task[s]" involved in mining (Holliday 1999, 272). Hence, the popular late-nineteenth-century saying, "You don't have a Chinaman's chance."

89. Jacobson 1984, 66.

90. The state of California numbers come from Kennedy 2000. The Santa Clara County statistics are from Keller 1981. On the whole, the California Native American population in 1848 outnumbered the white population by 10 to 1, but this was about to change rapidly (Rohrbough 1997, 8). The Native population in San Jose was further reduced to twelve persons by 1870, due mainly to smallpox and cholera (Heritage Media Corporation 1996, 79).

91. See *San Jose Mercury News,* May 4, 1992; see also Jacobson 1984. One of the other land grants to Ohlones was to a man named Marcelo. His grant, named Ulistac, now hosts several high-technology firms and Great America, the amusement park (see Jacobson 1984, 50).

92. Dasman 1999, 105.

93. Pisani 1999, 125. See also Jacobson 1984, 30.

94. Rohrbough 1997, 49.

95. So, for example, anyone who eats seafood today has some level of mercury in his or her body tissue. Mercury often originates from batteries, which are

frequently disposed of and burned in waste incinerators. The mercury then becomes airborne and settles in oceans where fish ingest it. Humans and other species then consume the fish and the toxins are passed on to us. For a particularly horrific example of the effects of mercury and methyl mercury on human health, consider the scourge of Minamata disease on a twentieth-century Japanese town. Mercury exposure can cause symptoms ranging from mild discomfort to nervous system damage and death (also see Lazaroff 2001a).

96. Holliday 1999, 253.

97. Kennedy 2000, 2001.

98. The San Francisco Regional Water Quality Control Board and public health officials have warned Bay Area communities not to consume fish caught in the Guadalupe River, as they may contain toxic mercury (see Stanley-Jones 2000).

99. St. Clair 1999, 198. Despite the fact that cyanide is now a well-known poison, it is the principal chemical used in gold mining around the world today. It is also widely used in Silicon Valley's electronics firms for the same purpose it served in 1890: to process gold. Only today, the gold is plated onto circuits and connectors and the cyanide is used as a reclaiming material for those metals. In January 1979, cyanide was among four chemicals involved in a chemical spill at Varian Associates, an electronics firm in Santa Clara. The spill took two and a half hours to contain; one employee was injured and the entire plant evacuated (see Malone and Yoachum 1980).

100. Ironically, hydraulic mining was supported as an occupational safety improvement over previous methods of mining, which were often accompanied by cave-ins from overhanging rocks and banks.

101. Holliday 1999, 251.

102. Pisani 1999, 133.

103. Rohrbough 1997, 203.

104. Heritage Media Corporation 1996, 73.

105. Keller 1981, 27.

106. Holliday 1999.

107. Roske 1968.

108. Jung 1999; Rohrbough 1997, 197.

109. Jung 1999.

110. Public Broadcasting Service 2001.

Notes to Chapter 3

1. The Spaniards were growing apricots as early as 1792 at the Mission Santa Clara de Asis, four miles northwest of San Jose. Daniel Post first introduced the canning business to California when he packed fruits, jellies, jams, and pickles beginning in 1854. The prune industry was established in the Santa Clara Valley

after the Frenchman Louis Pellier brought cuttings from the Agen district of France to San Jose in 1856 (see Torbert 1936, 248).

2. Dominguez 1992, 42.

3. Ibid., 44.

4. Heritage Media Corporation 1996, chap. 4.

5. McWilliams 1935 [1949], chap. 5.

6. Keller 1981, 30; San Jose Board of Trade 1887, 57. Alviso's low elevation and neglect by city planners would create problems, as evidenced in the floods of the winter of 1951–1952, when the entire town was evacuated (see Scott 1959, 278). Alviso's alluvial clay was land "thought to be inhospitable to any crop," so it is not surprising that "undesirables" like the Chinese, Japanese, and later Mexicanos were forced to live and work in this area (Lukes and Okihiro 1985, 12).

7. This pattern of ethnic succession, where new marginal groups were relegated to the jobs and neighborhoods of previous marginal groups, was common. In fact, Mexican immigrants and Chicanos moved into Alviso after the Japanese left years later.

8. Lukes and Okihiro 1985.

9. Examples include the strike by Chinese fruit pickers in August 1880 in the Santa Clara Valley and the sensational strike by Japanese workers at the Southern Pacific Railroad Company in 1904 (Lukes and Okihiro 1985).

10. On the successive waves of immigrants recruited for and then repelled from certain labor markets, McWilliams (1969, 134) writes, "it can be said that California has solved the difficult social problems involved in the use of alien labor by the simple expedient of driving the alien groups, one after the other, from the State."

11. McWilliams 1969, 112.

12. This managerial myth persists today and is regularly voiced by companies to justify majority female labor forces in Silicon Valley and in a range of exploitative industries throughout the world (see chapters 5, 6 and 7).

13. Bureau of Labor Statistics, State of California 1918, 28. Cf. Anthony 1928, 34.

14. Torbert 1936, 259. Like many immigrants from Asia and Latin America over the last one hundred fifty years, these Anglos, called "Arkies" and "Okies," were also driven to emigrate as the result of environmental problems. In fact, the Dust Bowl was one of the greatest human-caused ecological disasters in U.S. history.

15. U.S. Bureau of the Census 1973, Table 81.

16. U.S. Bureau of the Census 1980.

17. Working Partnerships USA and the Economic Policy Institute 1998, 13.

18. Daniels 1997.

19. McWilliams 1969, 134.

20. Dominguez 1992; Arnold 1976; Zavella 1987.

21. Heritage Media Corporation 1996, 160.

22. Chavez 1993. Pesticide poisoning is particularly acute for the hundreds of thousands of migrant children of color laboring in the nation's farmlands.

23. Brown 1981.

24. Anthony 1928, 14.

25. Burawoy 1979.

26. Anthony 1928, 46.

27. Ibid., 48.

28. Zavella 1987.

29. U.S. Department of Labor 1971, 361.

30. Department of Industrial Relations 1978, 7.

31. Zavella 1987, 108. Caustic and lye solutions are chemical mixtures used for loosening and removing the skin from vegetables and fruits (Jones 1973, 94, 98). They are quite hazardous, capable of burning, corroding, or destroying living tissue. Chlorine is used as a cleaning or disinfecting agent.

32. Zavella 1987, 108–109.

33. Ibid., 109.

34. Clark 1970.

35. Ibid., 83.

36. Ibid., 35.

37. Eighty-eight percent of the Valley's total acreage was farmland during the 1950s (see Clark 1970, 8).

38. Bade 1923, 178.

39. Matthews 1977, 32.

40. Torbert 1936, 247.

41. Heritage Media Corporation 1996, chap. 4.

42. While water was often scarce, there were times when residents had more than enough, with floods occurring in 1777, 1911, and 1951.

43. Jacobson 1984, 54–55.

44. Torbert 1936, 261.

45. Jacobson 1984, 55.

46. McWilliams 1935 [1949], 62–64. Pesticide spraying was also lucrative, with San Jose's FMC Corporation a major player in that market. The city's water works was privately owned by a firm on the East Coast, and this produced resentment among San Jose's citizenry. However, there was not enough political momentum to purchase the company from the owners. In fact, during the 1940s, one faction of political elites in San Jose argued that the water should be publicly owned, but they were quickly undermined by more powerful political forces—including the major newspaper, which dismissed them as socialists (see Trounstine and Christensen 1982, 88).

47. *San Jose Mercury-Herald*, February 13, 1941.

48. Hyde and Sullivan 1946. Raw sewage dumping directly into the bay produced a smell that was reputed to be the worst stench west of the Chicago stockyards (San Jose Planning Commission 1961, 47).

49. Alston 1990.

50. Heritage Media Corporation 1996, 212.

51. Sweeney 1997.

52. Ibid.

53. *San Jose News*, April 3, 1953.

54. Dominguez 1992.

55. *San Jose Mercury*, March 3, 1976.

56. Gruber 1976b.

57. A 1971 report documented that water pollution and solid waste pollution were the two biggest environmental problems related to the food manufacturing and processing industry (National Industrial Pollution Control Council 1971).

58. Townroe 1969.

59. Joachim 1979.

60. Matthews 1976, 68–69. A June 30, 1930, editorial in the *San Jose Mercury-Herald* asked rhetorically, "Why should we be escaping the effects of the business depression? Because of conservative growth, a good class of citizens, and good crops."

61. Matthews 1976, 178.

62. See *San Jose Mercury-Herald*, November 1 and November 4, 1937.

Notes to Chapter 4

1. In many ways, the history of FMC is the history of Silicon Valley. In 1883, in the village of Los Gatos, California (just south of San Jose), John Bean invented the first continuous flow agricultural pesticide spray pump. The next year he formed a small company to make pumps for nearby farmers. FMC went on to become the world's largest manufacturer of canning machines and a major player in the water/turbine pump business for the agricultural sector. Today FMC is a transnational manufacturer of machinery and chemicals used by electronics, military, agricultural, and timber industries around the world and has partnered with biotechnology firms as well (FMC 2001).

2. Stein 1973.

3. This dynamic of white flight and upward mobility into better-paying jobs occurred decades before, when Anglos and other European workers moved from the fields into the canneries in San Jose.

4. Keller 1981, 55.

5. Grass Roots Writing Collective 1970, 8–16.

6. Findlay 1992, 123.

7. Smith and Woodward 1992, 5.

8. Saxenian 1985a, 27.

9. *San Jose Mercury*, October 19, 1980.

10. Bernstein et al. 1977, 9; see also Choate 1968b.

11. *San Jose Mercury,* August 8, 1977; *San Jose News,* March 11, 1968.

12. Bernstein et al. 1977, 28.

13. A study by the county found that "the private sector has exerted a major influence in shaping the broad pattern of land development that has occurred in the County" (Santa Clara County Planning Committee 1973, 17).

14. Organization for Economic Cooperation and Development (OECD) 1968, 59–61, 63.

15. Findlay 1992, 132; David Packard, President of Stanford University Board of Trustees, quoted in *Daily Palo Alto Times,* February 17, 1960.

16. See Omi and Winant (1994) for an insightful discussion of the range of racial "code words" used as proxies for more direct words like "race" or "African Americans," "Asian Americans," etc.

17. Delgado and Stefancic 1999.

18. Choate 1968a.

19. Choate 1968c.

20. Fraser 1969.

21. Hanson 1969; Larimore 1969; Flood 1968.

22. Heritage Media Corporation 1996, 215.

23. The barrio of Sal si Puedes was home to many Chicanos. The name translates to "get out if you can" and the neighborhood earned this label because of the many mud holes in its streets before they were paved and because of the preponderance of substandard housing and poverty (see Clark 1970, 35).

24. Findlay 1992, 39; Clark 1970, vii.

25. Heritage Media Corporation 1996, 229. Recent research has also begun to make links among race, environmental quality, and transportation systems. Robert Bullard and his collaborators have traced the history of America's two-tiered, racially biased transportation system, from early conflicts over "separate but equal" accommodations, through "urban renewal," to today's urban sprawl that amplifies the problems of pollution and ecological damage (see Bullard and Johnson 1997; Bullard, Johnson, and Torres 2000).

26. Alviso 2001.

27. Rios 1987.

28. Dickey 1984.

29. This tension led to the Pentagon blocking Fujitsu Ltd., a Japanese corporation, from purchasing Fairchild Semiconductor Corporation in 1987, because of the alleged "national security risk" involved (see Sanger 1987). A senior White House staff member told a reporter, "This is a test case. If Japan can come in and buy this company, it can come in and buy them all over the place. We don't want to see the semiconductor industry under Japanese control" (Kilborn 1987). Pop-

ular bumper stickers in Santa Clara County, Detroit, and other areas where competition was fiercest included those reading: "Toyota, Datsun, Honda and Pearl Harbor" or "Hungry and Out of Work? Eat a Foreign Car." Never mind the fact that Fairchild Semiconductor Corporation was already controlled by the Schlumberger company, which was itself run by a French family; or that Intel Corporation was run by Andrew Gove, a Hungarian immigrant. One writer spoke of the "Japanese domination" of the 64K random access memory (RAM) chip in the 1980s as "the technological equivalent of Pearl Harbor" (Johnston 1982).

30. Pollack 1992.

31. Ibid.

32. IBM had a fifty-year-old tradition of no layoffs before it began letting employees go (via early retirements and buyouts) in the 1980s. When, in 1993, IBM announced that it would lay off employees for the first time in its history, they planned to cut up to seven thousand jobs, ending the fifty-year legacy of "full employment" adopted by IBM founder Tom Watson Sr. One computer peripheral manufacturer boasted in a classified ad for new workers about its "Recession Proof Strategy" (see *San Jose Mercury News*, February 1, 1982, 8CL). For more on layoffs, see "Electronics Job Losses Total 55,000 so far in '92," *San Jose Mercury News*, August 26, 1992; also see "IBM Ends No-Layoff Policy," *San Jose Mercury News Wire Services*, February 16, 1993.

33. Goldston 1984.

34. Marshall 1990.

35. Freedberg 2000.

36. See Park and Yoo 2001.

37. Goodell 2000.

38. Working Partnerships USA 1998, 13.

39. Ibid., 21–25.

40. Between 1990 and 1998, wages for the lowest-paid 25 percent of the Valley's workforce decreased by 14 percent (Sachs 1999, 18).

41. Siegel and Borock 1982, 36; Findlay 1992, 157.

42. United States Commission on Civil Rights 1982; see also Saxenian 1981.

43. Hossfeld 1988.

44. Siegel and Borock 1982, 23; Bernstein et al. 1977.

45. Heritage Media Corporation 1996, 226.

46. Trounstine and Christensen 1982, 92.

47. Ibid., 94.

48. Mandich 1975, 54.

49. Ibid., 92–94.

50. *San Jose Mercury*, February 6, 1952.

51. Mandich 1975, 65.

52. Cohen 1979.

53. Gruber 1976a.

54. See Sachs 1999, 19; Thurm 1993; Schalit 1984.
55. Cook and Thompson 2000.
56. Belser 1954.
57. Kent 1970, 30–34.
58. Whyte 1968, 54–56, 201.
59. Fox 1991; Hayes 1989.
60. Findlay 1992, 130–131.
61. Ironically, the SIP is now a federally designated toxic Superfund site, the result of leaking underground storage tanks and solvent spills by industrial tenants over the years (see Smith and Woodward 1992, 5).
62. Cummings 1982.
63. Page 2000, 40.
64. Ibid., 40.
65. Ibid., 43.
66. Kadetsky 1993, 519.
67. Industrial Survey Associates 1950, 4.
68. Scott 1959, 273.
69. Mandich (1975, 41) references these early boosters' claims of a pristine Valley during the 1950s and 1960s.
70. Siegel 1984, 58–59.
71. At that time, no toxicological study had proven a link between pollution and miscarriages or birth defects. One reason for this alleged lack of proof was the corruption and corporate bias endemic in the private laboratories and federal agencies conducting these tests (see Environmental Health Coalition 1992; Klinger 1982). For example, in one suspicious incident, of the 150 potentially toxic water samples the California Department of Health Services collected during the summer of 1980, the only one lost or destroyed was the sample from the water well Fairchild contaminated (Harris 1982). Perhaps CDHS officials were afraid of what they might find, particularly given the release of a report that very same summer revealing that the county cancer death rate had jumped 20 percent during the 1970s (see Klinger 1980). A more recent example of related corruption in environmental health testing was the indictment of employees of the Intertek Corporation for altering or falsifying some 250,000 environmental tests in 1996 and 1997. In many cases, false clean bills of health were issued for sites known to have significant environmental contamination, like the Rocky Mountain Arsenal in Denver, Colorado, and the Oakland Army Base in California (Ayres 2001).
72. Timberlake 1987, 142. Lorraine Ross and other activists in the Los Paseos neighborhood also discovered that toxicological knowledge about solvents was in its infancy during the mid-1970s, when the underground chemical storage tanks were built for Fairchild just two thousand feet from a public well (Yoachum 1982).

73. The group PHASE (Project on Health and Safety in Electronics) was at the forefront of this campaign. PHASE later changed its name to the Santa Clara Center for Occupational Safety and Health (SCCOSH), a group that continues to work closely with the Silicon Valley Toxics Coalition.
74. Timberlake 1987, 141.
75. SCCOSH 1982.
76. Champion 1988.
77. Benson 1986a.
78. Thurm 1992. Even more threatening than the contamination of the many smaller drinking wells was the pollution of the Anderson Reservoir—the county's prime drinking water source—by rocket fuel manufacturer United Technologies Corporation. In 1989, UTC spilled an estimated 10,000 gallons of groundwater contaminated with TCE into a creek that feeds into the reservoir (Calvert 1989).
79. *Hazardous Waste Contamination of Water Resources* 1985.
80. Kutzmann 1985.
81. Ibid.
82. Steinhart 1985.
83. Smith and Woodward 1992.
84. Johnston 1982, p.470.
85. Cummings 1982.
86. Hayhurst 1987.
87. Johnston 1982, 468.
88. Smith and Woodward 1992, 3.
89. Geiser 1985, pp. 11,45.
90. Rugemer 1987.
91. Sachs 1999.
92. Findlay 1992, 141.
93. Degnan 1968.
94. A January 25, 2000, *San Jose Mercury News* cartoon depicted a hamster in a cage, trying to sell the "property" over a cell phone: "How would I describe the place? . . . Cozy one bedroom, wood floors, large picture windows—a steal for under $400,000." According to an estimate provided by the local Emergency Housing Consortium, in 1997, approximately 20,000 employed, taxpaying citizens of Santa Clara County were forced to leave their homes and solicit beds at homeless shelters because of excessive rent hikes (Sachs 1999, 17).
95. Szasz and Meuser 2000.
96. *San Jose Mercury News*, April 28, 1988. Mountain View has one of the lowest median family incomes and highest percentages of people of color of any city in Santa Clara County (U.S. Bureau of the Census 1990).
97. Siegel 1984, 64.

98. With regard to solid waste, the white, affluent city of Palo Alto rejected all landfill proposals in 1976, thus shifting the burden of disposing of its own municipal solid waste to other communities.

99. Douglas 1987.

100. U.S. Department of Health and Human Services 1989a.

101. Lavelle and Coyle 1992.

102. Keller 1976. Two legally sanctioned dumps were located in the heavily African American, Latino, and Asian immigrant communities of Richmond and Martinez, north of Santa Clara County.

103. Personal communication with Jay Mendoza, director of the Health and Environmental Justice Project at SCCOSH and SVTC, May 2001. The source on asbestos and the landfills in Alviso is Allen-Taylor 1998.

104. While we must acknowledge that gaining Superfund status requires a good deal of political pressure on the part of communities, sometimes that status is successfully *resisted* by communities who view it as a stigma and a threat to their property values. For example, the communities of Breckenridge and Ward, Colorado, recently pressured the USEPA *not* to designate certain properties as Superfund sites.

105. Szasz and Meuser 2000, 607. TRI is the Toxics Release Inventory, a federal system of pollution tracking the USEPA uses to monitor large companies' environmental performance. While this program has a host of flaws, it is nevertheless a useful indicator of certain pollution trends in some large industries.

106. Stanley-Jones 1998.

107. Health and Environmental Justice Project 2001.

108. Bernstein et al. 1977, 19.

109. Scott 1959, 299.

110. See Mazurek 1999.

111. Trounstine and Christensen 1982. See Daly 1996a and b for a scientific explanation as to why an economic system predicated on infinite growth within a finite natural system is untenable.

Notes to Chapter 5

1. See Fox 1991; Gurtunya 1992; Hayes 1989; Hossfeld 1988; Siegel and Markoff 1985; and Sonnenfeld 1993.

2. We place the word "unskilled" in quotes because we believe no such thing exists as a job that requires zero skills. The term "unskilled" is an ideological construct that degrades the labor of the working classes and legitimates the unfair compensation white-collar workers receive.

3. See Gottlieb 1993, 2001; Pena 1997; Roberts 1993; Taylor 1997.

4. Sellers 1997.

5. One of the authors worked as an assembler in a cable-making firm in Sili-

con Valley during the winter and spring of 1999. Both authors conducted extensive interviews with workers and activists in the area and spent three years collecting documents from government, industry, community, and environmental advocacy sources. We were also granted access to many legal depositions with worker testimonies about toxic poisoning on the job.

6. Semiconductor Industry Association website: http://www.semichips.org.

7. *USA Today,* January 12, 1998.

8. Lazarus 2000.

9. Byster and Smith 1999.

10. "Have-Nots Struggling in Silicon Valley," *World-Herald,* December 20, 1998.

11. Ibid.

12. Data in this paragraph are from Working Partnerships USA 1998, 22, 25, 27.

13. Interview with Hector Guillen, San Jose, California, June 27, 2000.

14. *San Francisco Examiner,* April 25, 1993.

15. Hossfeld 1988; Park and Park 1999.

16. Park 1992, 231.

17. *Global Assembly Line* 1987.

18. Waller 1980.

19. Baker and Woodrow 1984, 22.

20. Conversation with Raquel Sancho of SCCOSH, October 8, 1999.

21. Here we are using "Asian" as an imprecise term encompassing Asian immigrants and Asian Americans, unless otherwise noted.

22. Gurtunya 1992, 86–87.

23. Johnson 1992.

24. Aragon and Goldman 1990.

25. Statement of Romulo Manan before the Santa Clara County Human Relations Commission. July 31, 1982.

26. Interview with Romi Manan, Milpitas, California, June 21, 2000.

27. Interview with Michael Eisenscher, Berkeley, California, June 2000.

28. Abate 1993.

29. Kay 1987.

30. Ibid.; see also Epstein 1979; Gottlieb 1993; Kazis and Grossman 1982; Pellow 2002; Sellers 1997; and Sheehan and Wedeen 1992 for research that supports this point.

31. Malone and Yoachum 1980.

32. Later, TCE was phased out in most firms and replaced with 1,1,1 trichloroethane (TCA), a substance that has since come under attack by environmentalists who suspect that it too is a harmful compound, as evidenced by the infamous Fairchild chemical spill in South San Jose *(San Jose Mercury News,* August 15, 1985).

33. Hawes Depositions 2000 (our emphasis).

34. Ted Smith at the June 27, 2000, WeLeaP! Graduation, San Jose, California.

35. SCCOSH 1995.

36. California Department of Health Services Press Release and CBDMP Information Packet, June 16, 1997. These illnesses may be the result of residential exposure, occupational exposure, or both.

37. Baker 1998.

38. Ted Smith, SVTC's Executive Director, has been one of the few visible male activists in the high-tech anti-toxics movement.

39. Melosi 1981.

40. By "feminist" we mean that these issues were addressed using the knowledge and experiences of women as both caretakers and professionals to operate these programs.

41. Malone and Yoachum 1980.

42. PHASE received sixty calls per week from concerned workers during one period and provided information to more than a thousand callers during the project's operations (Gurtunya 1992, 146).

43. Baker 1998.

44. See Churchill and Vanderwall (1989) for a wealth of documentation revealing the FBI's counterintelligence program aimed at destroying various radical organizations.

45. Baker 1998. Committees on Occupational Safety and Health (COSH) groups were founded in the 1970s by community, labor, and environmental organizations operating outside the auspices of trade unions. The purpose was to address worker concerns that were not being taken seriously by unions—namely health and safety (as opposed to just wages or "bread and butter" concerns) and issues that often intersected with environmental concerns (Kazis and Grossman 1982, 230–232).

46. Hawes Depositions 2000.

47. Dr. Jim Cone at the June 27, 2000, WeLeaP! Graduation, San Jose, California.

48. Interview with Raj Jayadev, October 9, 1999, Sonoma County, California.

49. Eisenscher 1993; Robinson and McIlwee 1989.

50. Interview with Michael Eisenscher, June 22, 2000, Berkeley, California.

51. We use the term "environmental justice" here to refer to the collection of social movement organizations committed to improving Silicon Valley working conditions, reducing toxics in the industry, empowering workers and residents in the Valley, and addressing these concerns at the state and industry levels.

52. Rios, Poje, and Detels 1993; Robinson 1991.

53. Gomes 1991.

54. *San Jose Mercury News*, February 7, 1993.

55. Levander 1992.

56. Benson 1987b.

57. Department of Industrial Relations, State of California 1978.

58. Wade et al. 1981.

59. Geiser 1985, 11; LaDou 1984. According to the U.S. Department of Health and Human Services, "Inorganic arsenic has been determined by NIOSH to be a carcinogen" (USDHHS 1987).

60. Pasguini and Laird 1982.

61. Cook and Thompson 2000.

62. Semiconductor Industry Association 1985.

63. Bureau of Labor Statistics 1986.

64. The electronics industry has never experienced a terribly high rate of physical injury. Instead, it is the rate of illness, often chemically related, that is so high.

65. Benson 1987b.

66. Gould, Schnaiberg, and Weinberg 1996; Hurley 1995.

67. This 1987 report by UC Davis researchers was a follow-up on the would-be investigation of the problem by the California Department of Industrial Relations, which was started and later suspiciously dropped when the political pressure against the investigation became too great (LaDou 1984). Thus, industry flexed its muscles and politicians and regulatory agencies buckled under pressure, despite the serious public health threats involved.

68. Silicon Valley Toxics Coalition 1999; also see chapter 4 of this book.

69. One explanation for these different priorities across agencies is that, in spite of the Reagan-Bush assault on both the EPA and OSHA (at the state and federal levels), the environmental movement was much more concerned about toxics than was the labor movement and consequently placed greater pressures on the appropriate agency—the EPA.

70. Benson 1986b.

71. Marshall 1987.

72. In 1985, the Federal Centers for Disease Control stated that human reproductive failure was "widespread and serious" and one of the ten most prevalent work-related diseases. Fetal Protection Policies (FPPs) are most common in lead, chemical, petrochemical, and oil refinery industries. Firms that reportedly have policies include Monsanto, Dow, DuPont, Union Carbide, B. F. Goodrich, and Allied Chemical. Many hospitals also have FPPs. A significant number of men who work in the manufacture of pesticides such as DBCP and Kepone have become sterile or fathered deformed children. The wives of men exposed to vinyl chloride in the workplace suffered a high incidence of stillbirths and miscarriages, according to another study. Finally, many male veterans of the Vietnam War believe the defoliant Agent Orange was the primary cause of their children's birth defects.

73. At the announcement of the DEC study, the pro-business publication *The Economist* was in rare form, displaying both its ignorance of electronics production and a stark disregard for women workers in toxic jobs: "As many as 1,000 different chemicals are used to make chips. Some of them are known to be dangerous. But most of the materials . . . are harmless. . . . Besides, the work is done in a clean room, that minimizes contact between people and chips. If this protects chips from people and their dust, clothing fibres and perfumes, it must also protect people from chips. Another possible—and simpler—reason for the high rate of miscarriages is that the workers spend so much time standing up" *(The Economist,* January 24, 1987, 82).

74. Schenker 1992. Also see Carlton 1992. A scientific review of the UC/SIA study later concluded that it appeared that the real cause of spontaneous abortions in clean rooms was exposure to glycol ethers in combination with xylene and n-butyl acetate (thus suggesting the necessity for understanding synergistic chemical effects). Other studies isolated fluorine and other chemicals as contributors to spontaneous abortions/miscarriages. For example, solvents such as isopropyl alcohol and acetone were believed responsible for increasing the risk of these types of health problems. Glycol ethers, it seems, are just one of at least seven chemical agents correlated with reproductive disorders.

75. Smith 1992a.

76. Ibid.

77. Butterfield 1992.

78. SCCOSH, SVTC, the Asian Workers Health Project, and the Campaign for Responsible Technology threatened this boycott (Smith 1992b).

79. The study, conducted by Johns Hopkins University, was commissioned in 1987 and begun in 1989 as a response to the Digital Equipment Corporation study.

80. Prior to the release of this study, some firms had already begun using ethyl lactate, a derivative of sugar beets, as a substitute for glycol ethers (Smith 1992a).

81. The finding that EGEs were indeed harmful to human health after years of use in electronics plants also lent credibility to the criticism by activists that the current regulatory system only has data on around 2 percent of the chemicals in current use—thus creating a "massive human experiment" wherein workers are the guinea pigs in a regime that views chemicals as "innocent until proven toxic." As Richard Wade, a Cal-OSHA official, stated to a reporter in 1980, "We have standards for about 2,000" of the 200,000 chemicals being used in the workplace at that time (Malone and Yoachum 1980).

82. Markoff 1992a and b.

83. Activists were repeatedly shut out of the Scientific Advisory Panel established by the SIA to review proposals for conducting the study. For example, in a November 1988 letter to the panel's chairperson, Dr. Patricia Buffler, national leaders of environmental and occupational health and safety organizations

pleaded with her to allow them to have input on the selection and advisory process, to no avail (Integrated Circuit Letter to Patricia Buffler, November 2, 1988, San Jose, California, on file at SCCOSH offices, San Jose, California).

84. Silicon Valley Toxics Coalition 1999.

85. *San Jose Mercury News,* February 23, 1998.

86. Ibid. A similar incident occurred at National Semiconductor, when that firm also ignored warnings of toxics and their health-related effects in the early 1980s (see chapter 8).

87. This process in high-tech industries is what Allan Schnaiberg and his collaborators have termed the "Treadmill of Production," a cycle of investment in ecologically disruptive technologies that creates more social and environmental harms, the solution to which is believed to lie in even more investment in the same or related systems of production (see Schnaiberg and Gould 1994; Gould, Schnaiberg, and Weinberg 1996; Weinberg, Pellow, and Schnaiberg 2000; and Schnaiberg 1994).

88. Silicon Valley Toxics Coalition 1998.

89. LaDou 1984.

90. Siegel, Smith, and Wilson 1990, 9–10.

91. Of course, for environmental and occupational health advocates, this is an argument for stricter regulation up front rather than experimenting on workers and later discovering that certain chemicals, or combinations thereof, are in fact harmful.

92. Harper 1986.

93. Ibid.

94. Gottlieb 1993.

95. Villones 1989. One of the problems with this relatively new and constantly changing industry is that the regulatory standards for acceptable levels of exposure to toxics are not always safe or reflective of reality. Examples abound. In May 1979, thirty-five workers at Verbatim Corporation of Sunnyvale reported headaches, nausea, and dizziness. However, Cal-OSHA could find no workplace cause, as the chemicals measured were well within regulatory limits, and dismissed the problem as "mass psychogenic illness" (Malone and Yoachum 1980; see also *USA Today,* January 12, 1998).

96. Lazarus 2000. Joseph LaDou, the occupational health expert who pioneered the focus on Silicon Valley workers beginning in 1968, laughed when told of Scalise's claim. "It constantly amazes me that people can call up the SIA and be told there's no problem. And they've been getting away with it for years" (ibid).

97. Alonso-Zaldivar 1998a.

98. *USA Today,* January 12, 1998.

99. Gurtunya 1992, 150. This particular claim is incredible, given the great number of high-tech workers whose asthma and allergies were caused or exacerbated by working conditions in the industry.

100. Begley and Carey 1984, 85.
101. Markoff 1992a.
102. Alonso-Zaldivar 1998a.
103. *San Jose Mercury News,* February 23, 1998.
104. Hawes Depositions 2000, 326–330.
105. Alonso-Zaldivar 1998b. This refusal to cooperate with health officials has happened in numerous cases in the United States and Japan (see earlier in this chapter and in chapter 8).
106. Welles 1984.
107. Hawes Depositions 2000, 287.
108. *Daily Camera,* July 4, 2001.

Notes to Chapter 6

1. Baker and Woodrow 1984, 24.
2. This worker's name is a pseudonym, as are all others unless they come from a previously published source.
3. Interview with Sarah Carpelli, June 2000, San Jose, California.
4. The Esperanza quotations in this paragraph are found in Hawes 2000, 12, 28, and 45.
5. Ibid., 59.
6. Roberts 1996.
7. Hawes 2000, 92.
8. Ibid., 134.
9. Interview with Sarah Carpelli, June 2000, San Jose, California.
10. Ibid.
11. Chu 1998.
12. Interview with Romi Manan, June 2000, Milpitas, California.
13. Interviews with Pat Lamborn of SCCOSH, February and November 1983.
14. Baker and Woodrow 1984, 25.
15. Ibid., 376.
16. Interview with Michael Eisenscher, Berkeley, California, June 2000.
17. Lamborn 1983.
18. Cone 1986.
19. Lazarus 2000.
20. Interview with Jose Carbon, October 10, 1999, San Francisco, California.
21. Hawes 2000, 127, 220.
22. Lamborn 1983.
23. Ibid., 218, 219.
24. Ibid., 355.
25. Ibid., pp. 99, 230-231.

26. Welles 1984.
27. Hawes 2000, 293, 300.
28. Lamborn 1983.
29. Gurtunya 1992, 149. Aside from this study, we have found no other evidence of this phenomenon, so it cannot be assumed that this practice was necessarily widespread.
30. Roberts 1996; Robinson 1991.
31. Burawoy 1979; Clawson 1980; Edwards 1979; Paules 1993.
32. Karasek and Theorell 1990; Roberts 1993; Robinson 1991.
33. Robinson and McIlwee 1989, 132.
34. Hawes 2000, 51.
35. Ibid., 64, 168, 206.
36. Ibid., 512–513.
37. Ibid., 108, 112.
38. Welles 1984.
39. Lamborn 1983.
40. Warren 1991.
41. See chapter 8 for a discussion of this phenomenon in Asia.
42. This quotation and the quotations in the following paragraph concerning Nancy Hawkes are from Malone and Yoachum 1980.
43. Lamborn 1983.
44. Ibid.
45. Ibid.
46. Interview with Romi Manan, June 2000, Milpitas, California.
47. In 1997, the USEPA concluded in its Design for Environment report on printed circuit/wire board toxics that "the most frustrating element in this study was its inability to fully evaluate all the chemicals in use by the industry or associated with new process technologies. This occurred because some companies that produce formulations used refused to reveal their chemical ingredients, claiming 'trade secrets.' The Project lacked authority to require them to divulge this information, having to rely on their voluntary good faith participation. Consequently, it is not possible to discount the possibility that some of the new process technologies may create hidden environmental or worker health hazards. . . . *Chemical companies should not be allowed to claim 'trade secrets' when worker and community health is at stake*, especially in an EPA-sponsored study" (Campaign for Responsible Technology 1997, 8, emphasis added).
48. This quotation and the material from Esperanza in the next paragraph are from Hawes Depositions 2000, 76, 98, 195, 208, 604–605.
49. The material from Lydia Johnson is from Hawes Depositions 2000, 303, 308, 438–441, 527.
50. *San Jose Mercury News*, February 23, 1998.
51. Institute of Medicine 1999b.

52. During a broadcast of National Public Radio in early May 2001, an internationally renowned cancer researcher stated, "it is now the consensus among the scientific community that cancer is fundamentally a genetic disease." This genetic orientation on cancer is a political, economically driven shift away from blaming corporate power and instead places the fault on the individual's genetic makeup.

53. See *Lancet* 1973; Epstein 1979, 77-79. Contributors who gave $100,000 or more to the American Cancer Society include carcinogen polluters General Electric and Dupont, and pharmaceutical companies Bristol-Meyers Squibb, Novartis, and Smith-Kline Beecham.

54. The W. R. Grace/Woburn case was featured in the book and movie *A Civil Action.* See also Brown and Mikkelsen (1990) for a strong sociological account of this tragedy.

55. *New York Times,* May 18, 2001. Examples of these conflicts of interest abound. American Cancer Society Board member David R. Bethune is president of Lederle Laboratories, a multinational pharmaceutical company and a division of American Cyanamid Company. Bethune is also vice president of American Cyanamid, which makes chemical fertilizers and pesticides—known carcinogens—while transforming itself into a full-fledged pharmaceutical company. In 1988, American Cyanamid introduced Novatrone, an anticancer drug. And in 1992, it announced that it would buy a majority of shares of Immunex, a cancer drug maker. The ACS is a large organization that provides research grants, public outreach, and expert advice on matters of public health relating to cancer.

56. Hawes 2000, 303–304.

57. Berman 1978; Kazis and Grossman 1982.

58. Begley and Carey 1984.

59. Hawes 2000, 199-200. Martino's main concerns were that the company was taking health information from her body without disclosing it to her and without telling her how it was to be used.

60. Hawes 2000, 202.

61. Ibid., 36.

62. Fisher 1998, emphasis added.

63. Johnson 1996, 7.

64. Mr. Prokopio 1996, 5.

65. Anonymous worker at WeLeaP! Graduation, December 6, 2000, San Jose, California.

66. Hawes 2000, 88.

67. Hawes 2000. Actual memo from the employee's physician to her employer.

68. Lamborn 1983.

69. Begley and Carey 1984.

70. Hawes 2000, 5.

Notes to Chapter 7

1. Kadetsky 1993, 517.
2. Ibid.
3. Working Partnerships USA 1998.
4. See chapters 4 and 5; Campaign for Responsible Technology 1997.
5. AIWA actually developed similar classes for women perhaps before SCCOSH, focusing on workplace literacy, leadership development, and health and safety education (see Kim and Kim 1996).
6. Raquel Sancho, SCCOSH Retreat, October 10, 2000.
7. Celia Chavez. WeLeaP! Worker Stories Process, 1999.
8. Anonymous Worker. Workers Stories Process at the SCCOSH WeLeaP! Graduation, December 6, 2000, San Jose, California.
9. Interview with Jung Sun Park, Cupertino, California, June 2000.
10. Ibid.
11. Raquel Sancho, SCCOSH Retreat, October 10, 2000.
12. In addition to winning his case with the California Labor Commissioner, Jayadev was featured in a documentary titled "The Secrets of Silicon Valley," released in 2001.
13. Jayadev 2001a.
14. Ibid.
15. Jayadev 2001a.
16. Excerpt from field notes, E-Tech, 1999.
17. Ibid.
18. Smith and Woodward 1992.
19. Karasek and Theorell 1990.
20. Gonos 1997.
21. Excerpt from Field notes, E-Tech, 1999. In fact, San Jose's living wage ordinance, passed in 1995, required city contractors to pay employees at least $10 per hour.
22. From a video produced for a temporary help firm in the San Francisco Bay Area, the particulars of which we are not at liberty to disclose.
23. Roberts 1996.
24. Excerpt from Field notes, E-Tech, 1999.
25. Fox's excellent study of GTE Lenkurt's plant in Albuquerque revealed this pattern too (Fox 1991, 12).
26. Excerpt from Field notes, E-Tech, 1999.
27. See Rollins 1985 and Salzinger 1997 for discussions of gendered deference rituals in several labor markets.
28. Ibid.
29. Excerpt from Field notes, E-Tech, 1999.
30. See Rollins (1985) for a stunning account of her undercover fieldwork as

a domestic servant in the Boston area. At one point, her employer and a family member held an intimate "private" conversation in the same room as Rollins (who was working as their domestic). Rollins felt so certain that they would pay her no attention that she boldly took out a notepad and pen and began to record field notes in their presence.

31. Southwick 1984.
32. Ibid.
33. Kelly Services 1999.
34. American Staffing Association 2000.
35. Hayes 1989, 175.
36. Rossi 2000.
37. Parker 1994, 47.
38. Jayadev 2001b.
39. Benner 1996.
40. Salcedo 2000.
41. Benner 1996.
42. Santa Clara Center for Occupational Safety and Health 2000a.
43. Interview with Korina, June 21, 2000, San Jose, California.
44. Manan 2000.
45. Cole-Gomolski 1999.
46. Greenhouse 2000a.
47. When we made this point during a presentation at a conference, an audience member asked, "In which country is your study based?" She was shocked to learn that these exploitative practices were not exclusive to the Third World, but were going on in the United States as well.
48. Carey and Malone 1980.
49. Markoff 1980.
50. Pellow and Park 2000.
51. Export processing zones are industrial areas in the Global South where manufacturers operate under special legal rights that are distinct from the normal laws of that nation. This includes labor, tax, and export regulations that allow for intensified exploitation of workers.
52. Christensen 1988a and b; Portes 1995, 31.
53. Lozano 1985.
54. This and the following quotations about home-based piecework are from Ewell and Ha 1999b.
55. Cook and Thompson 2000.
56. See Johnston 1982; Siegel and Borock 1982, 43–44.
57. Wahlin 2001, 7.
58. For example, see chapter 4 for a discussion of the Pentagon's obsession with Japanese high-tech competition during the 1980s.
59. Carey and Malone 1980.

60. Ewell and Ha 1999a.

61. Ewell and Ha 1999b.

62. Ewell and Ha 1999c.

63. Excerpt from Field notes, January 1999.

64. Interview with Victoria Rue, June 2000, Oakland, California.

65. Carey and Malone 1980.

66. As evidenced by the discovery of an even larger underground economy based on piecework in hundreds of people's homes, nearly twenty years later; and as evidenced by the Intel/USEPA Project XL fiasco (see chapter 10).

67. *San Jose Mercury News*, June 29, 1999.

68. Ewell 1999.

69. Ewell and Ha 1999c.

70. Ha 2000b.

71. SCCOSH and SBCLC 2000.

72. SCCOSH 2000b.

73. SCCOSH 2000c.

74. Wahlin 2001, 7.

75. Interview with Flora Chu, Palo Alto, California, June 2000.

76. Interview with K. Oanh Ha, San Jose, California, July 2000.

77. Erlich 1995.

78. Portes, Castells, and Benton 1989.

79. Glenn 1986. As Jung Sun, a Korean assembly worker, told us, "You just do it because you have to. You pull out every form of energy you have and focus on what needs to be done."

80. Furthermore, the evidence of unequal exposure in the workplace is much less susceptible to statistical manipulation and obfuscation than the studies of community-level exposure.

Notes to Chapter 8

1. Conference materials are from Arnold 2001. The absurdity of this idea was noted, ironically, by Microsoft's chairman, William H. Gates, who stated at the conference that the Internet was useless to people who lacked the basic necessities of life.

2. To take this analogy further, IBM Chief Executive Louis V. Gerstner was actually awarded an honorary British knighthood in 2001. He is the only American business leader to receive this award (Krane 2000).

3. Byster and Smith 1999.

4. http://www.svtc.org. "Globalization of High Tech."

5. Fernandez-Kelly and Schauffler 1994.

6. Military conflicts and global capital are deeply interconnected. For example, the electronics industry has for many years thrived precisely because of the

steady business from its military clients. Wars fought around the world in recent decades have been made possible by the microchip. Furthermore, as global capitalism produces scarcity of environmental and economic resources, military conflicts are one of the likely results.

7. Sassen 1998. Similar dynamics drive immigration into and migration between many of the world's major cities in both North and South.

8. For example, in Thailand, new factories and cities are powered by hydroelectric energy. One such project is the Bakun Dam, which has contributed to the destruction of nearby subsistence fishing and farming communities. As this local economy deteriorated, young people gravitated toward factories in the cities, depopulating their home villages. When unemployment hit, they had few options because their home economy and social structure had been crippled.

9. See chapter 1 for a discussion of the ecological dimensions of the anti-immigrant backlash in the United States.

10. See Armbruster, Bonacich, and Geron 1995; Chang 2000; Gwynne 1999; and Hondagneu-Sotelo 1994. Critiquing the classic "push-pull" theory of migration, Hondagneu-Sotelo (1994, 5) states that "the voluntarist assumptions embedded in this paradigm ignore the contingent social structural factors that shape migration, so that individual calculus occurs in a vacuum devoid of history and political economy." See also Saxenian 2000: "Researchers at the University of California at Berkeley have documented a significant correlation between the presence of first-generation immigrants from a given country and exports from California (for every 1 percent increase in the number of first-generation immigrants from a given country, exports from California to that country go up nearly 0.5 percent). Moreover, this effect is especially pronounced in the Asia-Pacific region where, all other things being equal, California exports nearly four times more than it exports to comparable countries in other parts of the world" (2000, 259).

11. Statement of Romulo Manan before the Santa Clara County Human Relations Commission, July 31, 1982. California Room, San Jose Public Library.

12. See Eisenscher 1993; Hossfeld 1988, forthcoming; Sevilla 1992.

13. Eisenscher 1993, 19–20.

14. Molina 1989, 145.

15. Ibid., 1.

16. Mazurek 1999, 89.

17. Personal interview, June 2000.

18. Still known for its basic assembly operations and contract manufacturing sector, Silicon Valley is now becoming famous for research and development in software and biotechnology.

19. Mazurek 1999, 109.

20. Henderson 1989, 51.

21. Gassert 1985, 39.

22. Interview, July 2000.

23. Ibid., 40, emphasis added. "Testing" is the final component in the semiconductor labor process.

24. For more detail on Japanese and U.S. competition, see Angel 1994.

25. Ibid., 40.

26. Ibid., 6. Because of the small size of the microchips, companies can load up a single airplane with several millions of dollars worth of these products and ship them overseas in a matter of hours.

27. Associated Press 2001.

28. Gassert 1985, 52.

29. See Gassert 1985; Hossfeld 1988, 1991; Nash and Fernandez-Kelly 1983.

30. In addition, export-oriented economic growth, measured by the Gross Domestic Product (GDP), is highly unstable and a distortion of reality. According to progressive economists Ted Halstead and Clifford Cobb, the use of the GDP as a measure of "progress" is "perverse" (Halstead and Cobb 1996). For example, the GDP does not take into account the depletion of natural resources; it also counts family breakdown, disease, public health crises, gun violence, and environmental catastrophes as economic gains and completely ignores transactions that are conducted without money.

31. Mazurek 1999, 9–10.

32. Gassert 1985, 5.

33. Byster and Smith 2001.

34. LaDou 1994, 767.

35. Semiconductor Industry Association (SIA) website: http://www.semichips.org.

36. Gassert 1985, 3.

37. Byster and Smith 2001.

38. This rule varies by city and is less stringent with regard to development in communities of color in the United States.

39. LaDou 1994, 767.

40. Kovaleski 1998.

41. Silicon Valley Toxics Coalition website: "Building an International Network." http://www.svtc.org/icrt/network.htm.

42. Byster 1997. Active opposition to the new Intel facilities in Costa Rica is evident. A new community organization, Asociacion Ecologica Belemita, has filed lawsuits to slow the fast-tracking of permits and semiconductor plant construction. Also, in January 1998, residents of Bosques de Dona Rosa, an affluent suburb in Costa Rica, clashed with police during a protest against the placement of power lines supplying the Intel plant. The residents claimed that the lines threatened their health and violated environmental regulations.

43. Lazarus 2000.

44. Morris 1998.

45. Lazarus 2000.

46. Chang 2000, 129. Chang also notes that approximately 700,000 Filipina/os were deployed through the Philippine Overseas Employment Administration each year during 1993 and 1994.

47. Sancho 1999.

48. Ibid.

49. Attorneys representing high-tech corporations against worker's compensation lawsuits often emphasize this difficulty as an important legal strategy. Our analysis of depositions from lawsuits brought by employees of several established high-tech firms supports revealed this as well.

50. Morris 1998. See chapter 9 for a more in-depth consideration of the Precautionary Principle.

51. Ibid.

52. Lazarus 2000.

53. See chapters 6 and 7.

54. This pattern of shifting toxic burdens from more powerful communities or world regions to less powerful areas is a form of environmental racism observed within the United States (see chapter 4) and between the United States and Africa, for example, with regard to hazardous waste disposal (Marbury 1995).

55. Summary quoted from Salazar n.d.

56. Daykin 1999; Elling 1981.

57. Lazarus 2000.

58. Gassert 1985, 4–5. This has long been a problem among workers in Silicon Valley as well (see chapters 5 and 6).

59. Gassert 1985, 26.

60. Information in this paragraph is from *The Economist,* July 25, 1998, 60.

61. Interview with Raquel Sancho of SCCOSH, October 18, 1998.

62. Gassert 1985, 64.

63. Ibid., 69.

64. Ibid.

65. Richards 1998.

66. Poling 1999. See chapter 5 for a discussion of glycol ethers in Silicon Valley and in New England high-tech industries.

67. Richards 1998.

68. Hossfeld 1988; LaDou 1994.

69. Gassert 1985.

70. Nash and Fernandez-Kelly 1983.

71. Baker and Woodrow 1984.

72. Richards 1998.

73. Hossfeld 1991.

74. Landler 2001.

75. Ibid.

76. Ibid.

77. Braun 1991; Gassert 1985, 44.

78. Khor 1996, 48.

79. Interestingly, this is the same general cause of an earlier regional industrial development policy, "import substitution industrialization," that imposed protectionist trade barriers to produce local and foreign markets in the 1950s and early 1960s. In this case, the lack of democratic reforms, corruption of local elites, greater class inequalities, and neglect of the agricultural sector impeded domestic economic sustainability (Gassert 1985, 45).

80. Baker 1984; Bello, Kinley, and Elison 1982.

81. Goldsmith 1996, 253.

82. Baker 1985. One such justification is the "fit hypothesis"—the idea that workers will naturally gravitate toward jobs that match their level of hazard and risk endurance. This becomes an academic justification for the stark correlations between race and hazardous occupations. Related to the fit hypothesis are the many myths and stereotypes managers use to argue that (immigrant) women are best suited for high-tech work because of their "nimble" and "fast" fingers and their allegedly cheerful disposition toward work (see chapters 3 and 5). A number of scholars have written on this subject. Hossfeld (1988) does a wonderful job pulling apart the gender, race, and class divisions of the high-tech international division of labor. Specific chapters in Sacks and Remy (1984) and Nash and Fernandez-Kelly (1983) are also noteworthy.

83. The example of Versatronex in chapter 9 supports this claim.

Notes to Chapter 9

1. "Ancestors in the Americas," PBS, March 30, 2001.

2. Scott 1959, 299.

3. Bernstein et al. 1977, 19.

4. For an in-depth analysis of a case of failed pollution prevention policy in the industry, see Mazurek 1999. Mazurek also offers a number of policy prescriptions for improvements in the high-tech sector.

5. We credit Allan Schnaiberg (1994) and his colleagues (Gould, Schnaiberg, and Weinberg 1996; Schnaiberg and Gould 1994; and Weinberg, Pellow, and Schnaiberg 2000) for this observation.

6. World Commission on Environment and Development 1987.

7. See DeSimone and Popoff 1997.

8. Ibid., 69.

9. Daly 1996b; Daly and Cobb 1994.

10. See Weinberg, Pellow, and Schnaiberg 2000.

11. For a defense of this position, see Peterson 1981. Much of the discipline of economics is based on this assumption as well (see Smith 1910).

12. Bryant 1995, 6.

13. Stanley-Jones 2001.

14. *Northern California Labor* 1984.

15. Lapin 1999.

16. See DeSimone and Popoff 1997.

17. In 1996, Monsanto's CEO, Robert Shapiro, wrote in the firm's *Environmental Review*, "Sustainable development will be a primary emphasis in everything we do." Cf. Bruno n.d.

18. See Semiconductor Industry Association website, http://www.semichips.org/esh/environmental.htm.

19. Sun Microsystems website, http://www.sun.com/aboutsun/ehs/ehs-design.html.

20. Apple Computer website, http://www.apple.com/about/environment//corporate/corporate.html.

21. Schnaiberg and Gould (1994) refer to this as the problem of "fully-only" social or environmental changes. That is, a representative of a chemical corporation might claim that their company has made "fully" X number of environmental improvements, while an environmentalist might counter that the firm has achieved "only" X number of initiatives. It is a "glass is half full" versus "glass is half empty" difference in perspectives.

22. Campaign member organizations include the Silicon Valley Toxics Coalition, the U.S. Public Interest Research Group, GrassRoots Recycling Network, Mercury Policy Project, and Clean Water Action.

23. *Silicon Valley Toxics Action* 19, no. 1 (spring 2001): 6.

24. Ibid. The Precautionary Principle is a concept that emerged during the late 1980s. It is also meant to constitute a challenge to the dominant method of hazard evaluation, "risk assessment." Risk assessment has long been critiqued by physicians, scientists, and activists for creating a situation where people and the environment are placed at higher than necessary risk; thus risk assessment is biased in favor of those institutions producing and profiting from these risks (i.e., corporations). Risk assessment is based on the assumption that there are always "acceptable exposure" levels to any chemical. This "risk paradigm" justifies placing toxics in our workplaces, neighborhoods, food, water, air, land, and bodies (see Montague 1998; see also O'Brien 2000 and Thornton 2000).

25. *Environment News Service,* April 16, 2001.

26. Lazaroff 2001b.

27. *Environment News Service,* April 16, 2001.

28. Baxter 1999.

29. Silicon Valley Toxics Coalition 2001b.

30. MacLachlan 1998; Thurm 1990.

31. Despite the EPA's claim that its voluntary initiative—the Global Warming Gas Emission Reduction Partnership—had an impact on these changes, it is more

likely that the 1987 Montreal Protocol (a global agreement to reduce greenhouse gases by 50 percent by 1998) and social movement campaigns have had a much greater impact. In fact, due partly to the phase-out of Freon (a CFC), in 1993 Silicon Valley firms produced less air pollution than in years past. It was also discovered that much of this drop in air pollution was due to companies shifting toxics to landfills and incinerators, and moving some of the dirtiest production to Third World nations (see Thurm 1993).

32. See Southwest Network for Environmental and Economic Justice and the Campaign for Responsible Technology 1997; *The Bargaining Chip* 1994, January.

33. The taxpayers' $100 million makes up half of SEMATECH's budget; the other half comes from the member companies.

34. Silicon Valley Toxics Coalition 1997; see also Smith 1992a.

35. The voluntary or "flexible" initiative was the Clinton administration's most popular approach to regulation. The USEPA and many national environmental organizations bought into the voluntary initiative approach as a way of avoiding confrontational tactics to bring industry into compliance with laws they regularly violate. Examples include the Common Sense Initiative, the EPA's WasteWi$e Program (designed to help companies reduce waste generation while maintaining profits), the EPA's Design for Environment Printed Wire Board Project (a voluntary partnership between the EPA and the Printed Wire Board trade association to reduce toxics and risks to workers in that sector), and the Global Warming Gas Emission Reduction Partnership—all of which have been near total failures for advancing the cause of environmental protection.

36. Smith 1995.

37. Cushman 1996; Hardie 1991.

38. Lewis 1997.

39. Most of this information is gleaned from the experience of one of the authors as an alternate representative to the CSI during the mid-1990s and from interviews the author conducted with environmentalists and occupational health advocates who participated in this project (see Pellow 1999b and c; see also Mazurek 1999).

40. Viederman 1998.

41. Which they have already done in Chicago.

42. Which has also already happened.

43. See Pellow 2002.

44. Interview with Raquel Sancho, October 8, 1998.

45. The language of "principles" has often been used by social movement organizations targeting various industries for reform. For example, the Valdez Principles (later named the CERES Principles)—designed to reform corporate environmental practices—emerged after the 1989 *Exxon Valdez* oil spill; the Precautionary Principle is aimed at all chemical industries, those industries using or

disposing of these chemicals, and regulators; and the Sullivan Principles were those proposed by anti-apartheid activists who demanded that U.S. companies operating in South Africa adopt affirmative action and civil rights–based practices in their South African offices and plants.

46. Good Neighbor Agreements are legally binding compacts between communities and corporations to ensure that the latter operates above regulatory standards vis-à-vis environmental and/or labor practices. These agreements usually emerge in the wake of industrial accidents that injure or kill workers and/or residents in fence-line communities (see Pellow 2001).

47. Silicon Valley Toxics Coalition, Campaign for Responsible Technology 1996.

48. Proposition 65 passed despite opposition from the San Jose City Council and most public agencies in the county, who feared that further regulation of the industry would be "ruinous to Silicon Valley's economy" (Robinson 1986).

49. *San Jose Mercury News,* July 13, 1990. Proposition 65 was a statewide, stronger version of the Federal Emergency Planning and Community Right to Know Act (EPCRA) that environmentalists helped pass that same year. These laws required companies with ten or more employees, handling at least 75,000 pounds of chemicals annually, to report the names and volumes of chemicals used and disposed of. The law covers more than 300 chemicals, and the reporting system—the Toxic Release Inventory (TRI)—gave ammunition to many environmentalists who publicly challenged numerous polluters for their toxics use. Some environmentalists used the TRI as the basis for lawsuits, organizing campaigns, and policy reform initiatives. For example, in 1988 SVTC issued a demand to the Valley's twenty-five biggest polluters to reduce by 90 percent the 12 million pounds of chemicals they dumped into the air and water annually (see Diringer 1988). The problems with this EPA-administered TRI system are many (see chapter 4), the first of which is that the polluters themselves are the ones doing the reporting—an honor system with no incentive to be honorable. This "fox guarding the chicken coop" scenario is troubling, particularly in light of Silicon Valley high-tech firms' propensity to lie and to suppress or withhold evidence when asked for data on environmental and occupational safety and health.

50. Conversation with Ted Smith, Executive Director of SVTC, December 2000.

51. In Seattle, for example, at-risk youth are taking computers out of the waste stream, learning to repair them, and selling them back to low-income Asian American/immigrant families. In the process, they are acquiring marketable skills. In Boulder, Colorado, the Colorado Materials Exchange has offered similar programs to urban youth and nonprofit organizations, to close the "digital divide" (Colorado Materials Exchange 2000). In Arlington, Virginia, the Electronics Industries Alliance recently unveiled an Internet-based plan—the Consumer Educa-

tion Initiative—that assists consumers who wish to reuse and recycle obsolete electronics products by giving them to charities or donating them to a school (*Environment News Service,* February 2, 2001). While these projects are laudable, the political and organizational challenges they face remain formidable. Foremost among them is the fact that recycling and reuse practices have steadily moved away from a socially conscious, volunteer-based movement run by activists to a market-based, profit-driven industry dominated by transnational corporations (Weinberg, Pellow, and Schnaiberg 2000). Thus, even seemingly progressive policy ideas are often quickly subsumed within the dominant global capitalist framework that values profits over people and the environment.

52. Robinson and McIlwee 1989.

53. Bose 1985.

54. Robinson and McIlwee 1989.

55. Cook 2000.

56. Zlolniski 1994.

57. Benner 1998.

58. Corporate Watch n.d.

59. Eng 1992b.

60. Eng 1992a.

61. DelBrocco 1992.

62. Smith 1992b.

63. See Rogers and Larsen 1984. The much-celebrated Internet-based book and music vendor Amazon.com recently faced a union organizing drive by the Communications Workers of America (CWA) in its Seattle offices. The firm quickly developed a counterintelligence campaign on its website, to "inform" workers that "unions actively foster distrust toward supervisors" and "also create an uncooperative attitude among associates by leading them to think they are 'untouchable' with a union" (Greenhouse 2000). One union advocate stated, "It's unfortunate that this vaunted high-tech company is just saying the same crude things that factory owners have been saying for 100 years about unions" (ibid.).

64. Interview with Michael Eisenscher, Berkeley, California, June 22, 2000. David Bacon was a National Semiconductor worker-activist who later became an accomplished photojournalist and member of the SCCOSH Board of Directors. Pat Lamborn and Mandy Hawes were SCCOSH founders.

65. Ibid.

66. Ibid.

67. Fan 2000.

68. Ibid.

69. Statement by anonymous government Official, San Jose, California, December 2000.

Notes to Chapter 10

1. For a refreshing critique of this postmodern vision of the world, see Huws 1999. For examples of the postmodernist vision of our era Huws critiques, see Cairncross 1997 and Coyle 1997.

2. Geiser 2001. See also Limerick 2000 for an analysis of this phenomenon as a challenge to Frederick Jackson Turner's thesis on the "frontier" in U.S. history.

3. Wayne 2001.

4. Accion Ecologica 1999

5. Huws 1999.

6. Mendoza 2001. *New York Times*, January 22, 2001. In 2001, the mayor of San Jose proposed to build sixteen power plants in various San Jose neighborhoods to alleviate the energy crisis. Not surprisingly, all but one of these locations were in communities of color. Specifically, fifteen of the sixteen neighborhoods had populations of color 67 percent or higher (see Health and Environmental Justice Project 2001). Supporting our contention that social movements are the main driver of progressive change in Silicon Valley, in July 2001 San Jose's mayor announced his intention to seek alternative locations for most of these power plants after EJ groups protested the plan, which they charged was an example of environmental racism. As a measure of its intense dependence on natural resources in this "postindustrial," "postmodern," and "Digital Age," the hike in energy prices during 2001 had Silicon Valley business screaming for relief and threatening to close down and move offshore where regulations are more "business friendly" (see *New York Times*, January 22, 2001).

7. Harden 2001, 35–36.

8. See Gedicks 1993.

9. Geiser 2001.

10. Braun 1991; Weinberg, Pellow, and Schnaiberg 2000.

11. Sassen 1998; Smith and Feagin 1987.

12. The European Union's Waste from Electrical and Electronic Equipment (WEEE) legislation is a prime example.

13. The canary in the coal mine metaphor refers to the practice, in previous centuries, of routinely lowering canaries into coal mines along with the miners. When the canary began to have trouble breathing or died, this alerted the miners that coal gas was present and they would evacuate.

14. This growing base of sufferers from environmental inequality is also a potential growing base of support for the EJ movement.

15. See Omi and Winant 1994 for a discussion of the foundations of U.S. sociology in the ethnicity school of thought, which is based on an underlying (and flawed) assumption that all ethnic groups can assimilate and blend into a sea of American-ness, if given the time.

16. For example, see Bryant and Mohai 1992; Bullard 2000; United Church of Christ 1987; United States General Accounting Office 1983.

17. Faber and Krieg 2001.

18. Yamane 1994.

19. Ibid.

20. Sayers 1994. It is our hope that this book will serve as a step toward creating "truth in history."

References

Abate, Tom. 1993. "Heavy Load for Silicon Valley Workers: Women, Minorities Face Uphill Battle as They Try to Climb Out of Low-Paying High-Tech Jobs." *San Francisco Examiner*, May 23.

Abbzug, Bonnie. 2001. "Shouldering Earth First!'s Baggage: Wilderness, Privilege and Immigration." *Earth First!* (May-June): 21.

Accion Ecologica. 1999. *Initial Proposal*. http://cosmovisiones.com/DeudaEcologica/c_propuin.html.

Adeola, Francis. 2000. "Cross-National Environmental Justice and Human Rights Issues—A Review of Evidence in the Developing World." *American Behavioral Scientist* 43: 686–706.

Allen-Taylor, J. Douglas. 1998. "Watchin' the Tidelands Roll Away." *MetroActive.com*, August 20-26.

Almaguer, Tomas. 1994. *Racial Fault Lines: The Historical Origins of White Supremacy in California*. Berkeley: University of California Press.

Alston, Dana. 1990. *We Speak for Ourselves*. Washington, D.C.: The Panos Institute.

Alonso-Zaldivar, Ricardo. 1998a. "High-Tech Industry Fights Study on Cancers." *Los Angeles Times*, December 6.

———. 1998b. "Industry Opposes Study of High-Tech Worker Cancers." *San Jose Mercury News*, December 6.

Alviso. 2001. "The Community of Alviso's Timeline." http://www.cachis.com/alviso/timeline.html.

American Staffing Association. 2000. "Staffing Facts." http://www.natss.org.

Amott, Teresa, and Julie Matthaei. 1991. *Race, Gender, and Work: A Multicultural Economic History of Women in the United States*. Boston: South End Press.

Anderton, Doug, A. B. Anderson, J. M. Oakes, and M. Fraser. 1994. "Environmental Equity: The Demographics of Dumping." *Demography* 31: 229–248.

Angel, David. 1994. *Restructuring for Innovation: The Remaking of the U.S. Semiconductor Industry*. New York: The Guilford Press.

Anthony, Donald. 1928. *Labor Conditions in the Canning Industry in the Santa Clara Valley of the State of California*. Ph.D. Thesis, Leland Stanford Junior University.

Aragon, Lawrence, and James Goldman. 1990. "Immigration Law May Boost Tech Firms." *The Business Journal*, November 12.

Armbruster, Ralph, Edna Bonacich, and Kim Geron. 1995. "The Assault on California's Latino Immigrants: The Politics of Proposition 187." *International Journal of Urban and Regional Research* 19: 655–663.

Arnold, Frank. 1976. "A History of Struggle: Organizing Cannery Workers in the Santa Clara Valley." *Southwest Economy and Society* 2: 26–38.

Arnold, Wayne. 2001. "Cell Phones for the World's Poor: Hook Up Rural Asia, Some Say, and Ease Poverty." *The New York Times*, January 19.

Asch, P., and J. L. Seneca. 1978. "Some Evidence on the Distribution of Air Quality." *Land Economics* 54: 278–297.

Associated Press. 2001. "Maxtor to Lay Off 700." *The New York Times*, June 21.

Ayres, Ed. 2001. "Lab Falsified Thousands of Tests." *World Watch*, March/April, 9.

Bade, William Frederic. 1923. *The Life and Letters of John Muir*. Volume 1. Boston: Houghton-Mifflin Company.

Baker, Dean. 1985. "The Study of Stress at Work." *Annual Review of Public Health* 6: 367–381.

Baker, Randall. 1984. "Protecting the Environment against the Poor." *Ecologist* 14, no. 2.

Baker, Robin. 1998. "Reflections of SCCOSH Founders." In *Silicon Dreams*. Santa Clara Center for Occupational Safety and Health's Workers' Memorial Day Book. May 3.

Baker, Robin, and Sharon Woodrow. 1984. "The Clean, Light Image of the Electronics Industry: Miracle or Mirage?" In Wendy Chavkin (ed.), *Double Exposure: Women's Health Hazards on the Job and at Home*. New York: Monthly Review Press.

Bargaining Chip: Bulletin of the Electronic Industry Good Neighbor Campaign. 1994. "Electronic Industry Good Neighbor Campaign." January.

Baxter, Kevin. 1999. "Environmental Groups Urge Tougher Recycling Rules." *American Metal Market*107, no. 98 (May 21): 7.

Been, Vicki. 1993. "Locally Undesirable Land Uses in Minority Neighborhoods." *Yale Law Review* 103: 1383.

Begley, Sharon, and John Carey. 1984. "Toxic Trouble in Silicon Valley." *Newsweek*, May 7, 85.

Bello, Walden, with David Kinley and Elaine Elison. 1982. *Development Debacle: The World Bank in the Philippines*. San Francisco: Institute for Food and Development Policy.

Belser, Karl. 1954. *Trends in Agricultural Conservation.* Santa Clara County Planning Department, December 10.

Benner, Chris. 1996. *Shock Absorbers in the Flexible Economy: The Rise of Contingent Employment in the Silicon Valley.* San Jose, Calif.: Working Partnerships USA.

———. 1998. "Win the Lottery and Organize: Traditional and Non-Traditional Labor Organizing in Silicon Valley." *Berkeley Planning Journal* 12: 50–71.

Benson, Mitchel. 1986a. "Boards Back Plan for IBM Toxic Monitors." *San Jose Mercury News,* March 19.

———. 1986b. "Health Officials Advised to Run Toxics Check on Valley Workers." *San Jose Mercury News*, September 27.

———. 1987a. "Santa Clara County's Killing Air." *San Jose Mercury News*, May 29.

———. 1987b. "Chip Worker Injuries Underreported." *San Jose Mercury News*, October 16.

Berman, Daniel. 1978. *Death on the Job: Occupational Health and Safety Struggles in the United States.* New York: Monthly Review Press.

Bernstein, A., DeGrrasse, B., Grossman, R., Paine, C., and Siegel, L. 1977. *Silicon Valley: Paradise or Paradox? The Impact of High Technology Industry on Santa Clara County*. Mountain View, Calif.: Pacific Studies Center.

Berry, B.J.L. (ed.). 1977. *The Social Burdens of Environmental Pollution: A Comparative Metropolitan Data Source*. Cambridge, Mass.: Ballinger Publishing Co.

Bose, Robin. 1985. "High-Tech Organizing: Unions Alone Are Not Enough." *The Guardian*, December 10.

Braun, Denny. 1991. *The Rich Get Richer: The Rise of Income Inequality in the United States and the World.* Chicago: Nelson-Hall Publishers.

Brimelow, Peter. 1995. *Alien Nation: Common Sense about America's Immigration Disaster.* New York: Random House.

Broek, J. M. 1932. *The Santa Clara Valley: A Study in Landscape Changes.* Utrecht, The Netherlands: N.V.A. Oosthoek's Uitgevers.

Brown, Martin. 1981. *A Historical Economic Analysis of the Wage Structure of the California Fruit and Vegetable Canning Industry*. Ph.D. dissertation, University of California, Berkeley.

Brown, Phil, and Faith Ferguson. 1995. "'Making a Big Stink': Women"s Work, Women"s Relationships, and Toxic Waste Activism." *Gender & Society* 9:145–172.

Brown, Phil, and Edwin Mikkelsen. 1990. *No Safe Place: Toxic Waste, Leukemia and Community Action*. Berkeley: University of California Press.

Bruno, Kenny. n.d. "Greenwash Award of the Month: Monsanto." Corporate Watch Website, http://www.corpwatch.org.

Bryant, Bunyan, and Paul Mohai. 1992. *Race and the Incidence of Environmental Hazards: A Time for Discourse*. Boulder, Colo.: Westview Press.

Bryant, Bunyan (ed.). 1995. *Environmental Justice: Issues, Policies, and Solutions*. Washington, D.C.: Island Press.

Bullard, Robert. 2000. *Dumping in Dixie: Race, Class and Environmental Quality*. Boulder, Colo.: Westview Press.

Bullard, Robert (ed). 1993. *Confronting Environmental Racism: Voices from the Grassroots*. Boston: South End Press.

———. 1994. *Unequal Protection*. San Francisco: Sierra Club Books.

Bullard, Robert, Eugene Grigsby, and Charles Lee (eds.). 1994. *Residential Apartheid: The American Legacy*. Los Angeles: CAAS Publishers.

Bullard, Robert, and Glenn Johnson (eds.) 1997. *Just Transportation: Dismantling Race and Class Barriers to Mobility*. Philadelphia: New Society Publishers.

Bullard, Robert, Glenn Johnson, and Angel Torres (eds.). 2000. *Sprawl City: Race, Politics, & Planning in Atlanta*. San Francisco: Island Press.

Bullard, Robert, and Beverly Wright. 1993. "The Effects of Occupational Injury, Illness and Disease on the Health Status of Black Americans." In Richard Hofrichter (ed.), *Toxic Struggles: The Theory and Practice of Environmental Justice,* 153–162. Philadelphia: New Society Publishers.

Burawoy, Michael. 1979. *Manufacturing Consent: Changes in the Labor Process under Monopoly Capitalism*. Chicago: University of Chicago Press.

Bureau of Labor Statistics, State of California. 1918. *Labor Conditions in the Canning Industry.*

Bureau of Labor Statistics. 1986. *Occupational Injuries and Illnesses in the United States by Industry, 1984*. U.S. Department of Labor, Bulletin 2259, May.

Bustamante, Jorge. 1997. "Mexico–United States Labor Migration Flows." *International Migration Review* 31: 1112–1121.

Butterfield, Bruce. 1992. "Groups Hit Chip Industry's Policies." *The Boston Globe*, January 16.

Byster, Leslie. 1997. "Intel Inside Costa Rica." *Silicon Valley Toxics Action Newsletter*, Silicon Valley Toxics Coalition.

Byster, Leslie, and Ted Smith. 1999. "High-Tech and Toxic: Global Labor Pains in the Electronics Industry." *Forum for Applied Research and Public Policy* 14, no. 1 (spring): 69–76.

———. 2001. "Uncovering Problems and Promise in Taiwan." *Silicon Valley Toxics Action Newsletter* 14, no. 1.

Caddes, Carolyn. 1986. *Portraits of Success: Impressions of Silicon Valley Pioneers*. Palo Alto, Calif.: Tioga Publishing.

Cairncross, Frances. 1997. *The Death of Distance: How the Communications Revolution Will Change Our Lives*. Boston: Harvard Business School Press.

California Department of Health Services Press Release and CBDMP Information Packet. June 16, 1997.

Calvert, Cathie. 1989. "Environmentalists Rap UTC: Quick Cleanup of Toxics Asked." *San Jose Mercury News*, November 8.
Camacho, David E. 1998. *Environmental Injustices, Political Struggles: Race, Class, and the Environment*. Durham, N.C.: Duke University Press.
Campaign for Responsible Technology. 1997. *EPA Report on Printed Wire Board Technologies*. March.
Carey, Peter. 1986. "Introduction: The Capital of the Future." In Carolyn Caddes, *Portraits of Success: Impressions of Silicon Valley Pioneers*. Palo Alto, Calif.: Tioga Publishing Company.
Carey, Pete, and Michael Malone. 1980. "Black Market in Silicon Valley." *San Jose Mercury*, August 31.
Carlton, Jim. 1992. "Computer-Chip Worker Study Confirms Risk of Miscarriages." *Wall Street Journal*, December 4.
Castillo, Edward. 1989. "An Indian Account of the Decline and Collapse of Mexico's Hegemony over the Missionized Indians of California." *American Indian Quarterly* 13: 399.
———. 1994. "The Language of Race Hatred." In Lowell John Bean (ed.), *The Ohlone: Past and Present: Native Americans of the San Francisco Bay Region*, 272–295. Menlo Park, Calif.: Ballena Press.
Castleman, Barry. 1987. "Workplace Health Standards and Multinational Corporations in Developing Countries." In Charles Pearson (ed.), *Multinational Corporations, Environment, and the Third World*, 149–174. Durham, N.C.: Duke University Press.
Champion, Dale. 1988. "Plan for Cleaning Silicon Valley Water." *San Francisco Chronicle*, October 20.
Chan, Sucheng. 1986. *This Bittersweet Soil: The Chinese in California Agriculture, 1860–1910*. Berkeley: University of California Press.
———. 2001. *Ancestors in the Americas*. Public Broadcasting Service, March 30.
Chang, Grace. 2000. *Disposable Domestics: Immigrant Women Workers in the Global Economy*. Boston: South End Press.
Chavez, Cesar. 1993. "Farm Workers at Risk." In Richard Hofrichter (ed.), *Toxic Struggles: The Theory and Practice of Environmental Justice*, 163–170. Philadelphia: New Society Publishers.
Chavis, Ben. 1993. "Environmental Racism." In *Confronting Environmental Racism: Voices from the Grassroots*, ed. Robert Bullard, 1–8. Boston: South End Press.
Choate, Jim. 1968a. "Valley Jolted Out of Lethargy." *San Jose News*, March 7.
———. 1968b. "Brains Plus Technology Equals 'Boom.'" *San Jose News*, March 9.
———. 1968c. "Ours Is Truly a White-Collar County." *San Jose News*, March 15.

Christensen, Kathleen E. 1988a. *The New Era of Home-Based Work: Directions and Policies*. Boulder, Colo.: Westview Press.

———. 1988b. *Women and Home-based Work: The Unspoken Contract.* New York: Henry Holt and Co.

Chu, Flora. 1998. "Asian Workers Health Project." In *Silicon Dreams*. Santa Clara Center for Occupational Safety and Health's Workers' Memorial Day Book. May 3.

Churchill, Ward. 1993. *Struggle for the Land: Indigenous Resistance to Genocide, Ecocide and Expropriation in Contemporary North America*. Monroe, Maine: Common Courage Press.

Churchill, Ward, and Jim Vanderwall. 1989. *Agents of Repression: The FBI's Secret Wars against the Black Panther Party and the American Indian Movement*. Boston: South End Press.

Clark, Margaret. 1970. *Health in the Mexican-American Culture*. Berkeley: University of California Press.

Clark, Thomas (ed.). *Gold Rush Diary.* Lexington: University of Kentucky Press.

Clawson, Dan. 1980. *Bureaucracy and the Labor Process: The Transformation of U.S. Industry, 1860–1920*. New York: Monthly Review Press.

Clune, Michael. 1998. "The Fiscal Impacts of Immigrants: A California Case Study." In James P. Smith and Barry Edmonston (eds.), *The Immigration Debate: Studies of the Economic, Demographic, and Fiscal Effects of Immigration*, 120–182. Washington, D.C.: National Academy Press.

Cohen, Susan. 1979. "The Hunt for Land." *San Jose Mercury News*, January 28.

Cole, Luke. 1992. "Empowerment as the Key to Environmental Protection: The Need for Environmental Poverty Law." *Ecology Law Quarterly* 19: 619.

Cole, Luke, and Sheila Foster. 2001. *From the Ground Up: Environmental Racism and the Rise of the Environmental Justice Movement*. New York: New York University Press.

Cole-Gomolski, Barb. 1999. "Court Orders Benefits for Microsoft Temps." *Computerworld*, May 13.

Colorado Materials Exchange. 2000. *Making the Most of Colorado's Materials*. Boulder, Colo.: University of Colorado Recycling Services.

Communications Workers of America, AFL-CIO, CLC. 1991. *Occupational Safety & Health Fact Sheet #23: High Tech Toxics & the Workplace*. Washington, D.C.: AFL-CIO.

Cone, James. 1986. "Health Hazards of Solvents." *State of the Art Reviews: Occupational Medicine* 1, no. 1 (January-March): 69-87.

Cook, Christopher. 2000. "Temps Demand a New Deal." *The Nation*, March 27.

Cook, Christopher, and Clay Thompson. 2000. "Silicon Hell." *San Francisco Bay Guardian*, April 26, 16–27.

Cook, Sherburne F. 1951. "Antiquity of San Francisco Bay Shellmounds." In Robert F. Heizer and M. A. Whipple (eds.), *The California Indians: A Source Book*, 172–175. Berkeley: University of California Press.

———. 1976. *The Conflict between the California Indians and White Civilization*. Berkeley: University of California Press.

Cornford, Daniel. 1999. "'We All Live More Like Brutes than Humans': Labor and Capital in the Gold Rush." In James Rawls and Richard Orsi (eds.), *A Golden State: Mining and Economic Development in Gold Rush California*, 78–104. Berkeley: University of California Press.

Corporate Watch. n.d. "Organizing the High Tech Industry." http://www.corpwatch.org.

Coursey, Don. 1994. "Environmental Racism in the City of Chicago." Paper presented at the Irving B. Harris School of Public Policy, University of Chicago. October.

Coyle, Diane. 1997. *Weightless World: Strategies for Managing the Digital Economy*. Oxford: Capstone Publishing.

Cummings, Judith. 1982. "Leaking Chemicals in California's 'Silicon Valley' Alarm Neighbors." *New York Times*, May 20.

Cushman, John. 1996. "EPA and Arizona Factory Agree on Innovative Regulatory Plan." *New York Times*, November 20, A18.

Boulder (Colo.) Daily Camera. 2001. "IBM Trims 1,500 Jobs." July 4.

Daly, Herman E. 1996a. "Sustainable Growth? No Thank You." In Jerry Mander and Edward Goldsmith (eds.), *The Case against the Global Economy*. San Francisco: Sierra Club Books.

———. 1996b. *Beyond Growth: The Economics of Sustainable Development*. Boston: Beacon Press.

Daly, Herman E., and John B. Cobb. 1994. *For the Common Good*. Boston: Beacon Press.

Daniels, Roger. 1997. *Not Like Us: Immigrants and Minorities in America, 1890–1924*. Chicago: Ivan R. Dee.

———. 1997. "United States Policy towards Asian Immigrants." In Darrell Y. Hamamoto and Rodolfo D. Torres, *New American Destinies: A Reader in Contemporary Asian and Latino Immigration*. New York: Routledge.

Dasman, Raymond. 1999. "Environmental Changes Before and After the Gold Rush." In James Rawls and Richard Orsi (eds.), *A Golden State: Mining and Economic Development in Gold Rush California*, 105–122. Berkeley: University of California Press.

Daykin, Norma. 1999. "Introduction: Critical Perspectives on Health and Work." In Norma Daykin and Lesley Doyal (eds.), *Health and Work: Critical Perspectives*. New York: St. Martin's Press.

Daykin, Norma, and Lesley Doyal (eds.). 1999. *Health and Work: Critical Perspectives*. New York: St. Martin's Press.

DeFreitas, Gregory. 1995. "Immigration, Inequality, and Policy Alternatives." Discussion paper presented at the Conference on Globalization and Progressive Economic Policy, Economic Policy Institute, Washington, D.C., October 28.

Degnan, James P. 1968. "Santa Cruz: A Workable Utopia." In Carey McWilliams (ed.), *The California Revolution*. New York: Grossman Publishers.

DelBrocco, Pete. 1992. "Versatronex Says Safety Is Top Priority." *San Jose Mercury News,* November 23.

Delgado, Richard, and Jean Stefancic. 1999. "Home-Grown Racism in Colorado." Paper presented at Old main Auditorium, University of Colorado, October.

Department of Agricultural Engineering. 1977. *Research in Cannery Noise Control*. Davis: University of California, Davis.

Department of Industrial Relations. 1978. *Occupational Injuries and Illnesses Survey, California, 1976*. Sacramento: State of California, Division of Labor Statistics and Research.

DeSimone, Livio, and Frank Popoff. 1997. *Eco-Efficiency: The Business Link to Sustainable Development*. Cambridge, Mass.: MIT Press.

Diaz, David. 2001. "Environmental Logic and Minority Communities." In Marta Lopez-Garza and David R. Diaz (eds.), *Latino Immigrants in a Restructuring Economy: The Metamorphosis of Southern California*, 425–447. Stanford, Calif.: Stanford University Press.

Dickey, Jim. 1984. "Anti-Asian Bigotry on Rise." *San Jose Mercury News*, February 4.

Diringer, Elliot. 1988. "Foes of Toxics Sic New Law on Silicon Valley Firms." *San Francisco Chronicle*, August 3.

Dittgen, Herbert. 1997. "The American Debate about Immigration in the 1990s: A New Nationalism After the End of the Cold War?" *Stanford Humanities Review* 5, no. 2: 256–286.

Dominguez, RoseAnn. 1992. *The Decline of Santa Clara County's Fruit and Vegetable Canning Industry (1967-1987)*. Master's thesis, Department of Geography, San Jose State University.

Douglas, Sally. 1987. "Lorentz Jailed; Clean-up Stalled." *Silicon Valley Toxics News* (summer).

D'Souza, Karen. 2000. "Giving Voice to Valley's Have-Nots." *San Jose Mercury News,* April 20.

Economist, The. 1987. "Health Hazards at Work—The Risks of the Clean Room." January 24, 82.

———. 1998. "Japanese Electronics Found to Pollute Groundwater." July 25, 60.

Edwards, Richard. 1979. *Contested Terrain: The Transformation of the Workplace in the Twentieth Century*. New York: Basic Books.

Equal Employment Opportunity Commission (EEOC). 1980. *EEOC Summary Report of Selected Establishments from the Technical Services Division.* Equal Opportunity Commission.

Eisenscher, Michael. 1993. "Silicon Fist in a Velvet Glove." Unpublished paper, November.

Elling, R. H. 1981. "Industrialisation and Occupational Health in Underdeveloped Countries." In Vicente Navarro and Daniel Berman (eds.), *Health and Work under Capitalism: An International Perspective.* New York: Baywood.

Eng, Sherri. 1992a. "Client Feels Heat in Versatronex Labor Dispute." *San Jose Mercury News*, November 13.

———. 1992b. "Striking Versatronex Workers End 6-Week Walkout." *San Jose Mercury News*, December 2.

Environment. 1999. "High Dreck." June, 22.

Environment News Service. 2001a. "Website Helps Consumers Recycle Used Electronics." February 2.

———. 2001b. "Electronics Makers Pressed to Take Back Discards." April 16.

———. 2001c. "Living Free of Pollution Called Basic Human Right." April 30.

Environmental Health Coalition. 1992. *Inconclusive By Design.* San Diego, Calif.: EHC.

Epstein, Samuel S. 1979. *The Politics of Cancer.* Garden City, N.Y.: Anchor Books.

Erikson, Kai. 1994. *A New Species of Trouble.* New York: W. W. Norton.

Erlich, Reese. 1995. "Prison Labor: Workin' for the Man." *Covert Action Quarterly* 54 (fall).

Espenshade, Thomas J., and Maryann Belanger. 1997. "U.S. Public Perceptions and Reactions to Mexican Migration." In Frank Bean, Rodolfo O. de la Garza, Bryan Roberts, and Sidney Weintraub (eds.), *At the Crossroads: Mexican Migration and U.S. Policy,* 227–261. New York: Rowman & Littlefield.

Ewell, Miranda. 1999. "Agencies Probing Piecework." *San Jose Mercury News*, July 8.

Ewell, Miranda, and Oanh Ha. 1999a. "High Tech's Hidden Labor." *San Jose Mercury News,* June 26.

———. 1999b. "Long Nights and Low Wages." *San Jose Mercury News*, June 27.

———. 1999c. "Piecework Diminishes." *San Jose Mercury News*, October 17.

Faber, Daniel, and Eric Krieg. 2001. *Unequal Exposure to Ecological Hazards: Environmental Injustices in the Commonwealth of Massachusetts.* A Report by the Philanthropy and Environmental Justice Research Project, Northeastern University, Boston, Massachusetts.

Fan, Maureen. 2000. "Unions' Latest Organizing Push: Immigrants." *San Jose Mercury News*, January 21.

Feagin, Joe, and Melvin P. Sikes. 1994. *Living with Racism: The Black Middle-Class Experience*. Boston: Beacon Press.
Fellmeth, Robert C. (ed.). 1973. *Politics of Land: Ralph Nader's Study Group Report on Land Use in California*. New York: Grossman Publishers.
Fernandez-Kelly, Maria Patricia. 1983. *For We Are Sold, I and My People: Women and Industry in Mexico's Frontier.* Albany: State University of New York Press.
Fernandez-Kelly, Mary Patricia, and Richard Schauffler. 1994. "Divided Fates: Immigrant Children in a Restructured U.S. Economy." *International Migration Review* 28: 662–689.
Findlay, John M. 1992. *Magic Lands: Western Cityscapes and American Culture after 1940*. Berkeley: University of California Press.
Fisher, Tom. 1998. "Learning the Hard Way: Injured Workers United." In *Silicon Dreams*. Santa Clara Center for Occupational Safety and Health's Workers' Memorial Day Book. May 3.
Flood, Dick. 1968. "EOC Meeting 'Erupts Like Volcano.'" *San Jose News*, March 14.
Food Machinery and Chemical Corporation. 2001. FMC Machinery and Chemicals. http://www.fmc.com.
Foner, Nancy. 1998. "The Transnationals." *Natural History* 107: 34–35.
Fox, Steve. 1991. *Toxic Work: Women Workers at GTE Lenkurt.* Philadelphia: Temple University Press.
Fraser, Jack. 1969. "Militant Mexican-Americans on Rise." *The Mercury*, January 11.
Freedberg, Louis. 2000. "Labor Shortage Is Turning the Anti-Immigrant Tide." *San Francisco Chronicle*, March 12.
Freeman, A. M. III. 1972. "The Distribution of Environmental Quality." In A. V. Kneese and B. T. Bower (eds), *Environmental Quality Analysis: Theory and Method in the Social Sciences*, 243–278. Baltimore, Md.: Published for Resources for the Future by Johns Hopkins University Press.
Friedman-Jimenez, George. 1994. "Achieving Environmental Justice: The Role of Occupational Health." *Fordham Urban Law Journal* (spring).
Fujiwara, Lynn. 1998. "The Impact of Welfare Reform on Asian Immigrant Communities." *Social Justice* 25:82–103.
Galvin, John (ed.). 1971. *The First Spanish Entry into San Francisco Bay, 1775.* San Francisco: John Howell.
Gassert, Thomas H. 1985. *Health Hazards in Electronics—A Handbook*. Hong Kong: Asia Monitor Resource Center.
Gedicks, Al. 1993. *The New Resource Wars: Native and Environmental Struggles against Multinational Corporations*. Boston: South End Press.
Geiger, Maynard. 1951. "The Arrival of the Franciscans in the Californias—1768–1769, According to the Version of Fray Juan Crespi, O.F.M." *The Americas* 8: 209–218.

Geiser, Ken. 1985. "The Chips Are Falling: Health Hazards in the Microelectronics Industry." *Science for the People* 17: 8.

———. 2001. *Materials Matter: Toward a Sustainable Materials Policy.* Cambridge, Mass.: MIT Press.

Gillonna, John. 2000. "San Francisco Dot-Coms Blasted for Changing City." *Boulder (Colo.) Daily Camera,* October 4.

Glenn, Evelyn Nakano. 1986. *Issei, Nissei, Warbride.* Philadelphia: Temple University Press.

———. 1992. "From Servitude to Service Work: Historical Continuities in the Racial Division of Paid Reproductive Labor." *Signs: Journal of Women in Culture and Society* 18: 1–43.

Global Assembly Line. 1987. Documentary Film. Newday Filsm. 58 minutes.

Goldsmith, Edward. 1996. "Development as Colonialism." In Jerry Mander and Edward Goldsmith (eds.), *The Case against the Global Economy,* 253–266. San Francisco: Sierra Club Books.

Goldston, Linda. 1984. "INS: 25% of Workers Here Illegally." *San Jose Mercury News,* April 17.

Gomes, Lee. 1991. "Haves Have Computers, Have-nots Don't." *San Jose Mercury News,* March 27.

Gonos, George. 1997. "The Contest over 'Employer' Status in the Postwar United States: The Case of Temporary Help Firms." *Law and Society Review* 31: 81–110.

Goodell, Jeff. 2000. "The Venture Capitalist in My Bedroom." *The New York Times Magazine,* May 28, 32–59.

Gottlieb, Robert. 1993. *Forcing the Spring: The Transformation of the American Environmental Movement.* Washington, D.C.: Island Press.

———. 2001. *Environmentalism Unbound: Exploring New Pathways for Change.* Cambridge, Mass.: MIT Press.

Gould, Kenneth, Allan Schnaiberg, and Adam Weinberg. 1996. *Local Environmental Struggles: Citizen Activism in the Treadmill of Production.* New York: Cambridge University Press.

Grass Roots Writing Collective. 1970. *The Promised Land: A Grass Roots Report on Mid-Peninsula Land Use.* Palo Alto, Calif.: Stanford Chapparal.

Green, Susan S. 1983. "Silicon Valley's Women Workers: A Theoretical Analysis of Sex-Segregation in the Electronics Industry." In June Nash and Maria Patricia Fernandez-Kelly (eds.), *Women, Men and the International Division of Labor.* Albany: State University of New York Press.

Greenhouse, Steven. 2000a. "Labor Board Makes Union Membership Easier for Temps." *New York Times,* August 31.

———. 2000b. "Amazon.com Is Using the Web to Block Unions' Effort to Organize." *New York Times,* November 29.

Gruber, Stephen. 1976a. "Sewage, Electric Plant Gets OK." *San Jose Mercury News*, February 2.
———. 1976b. "Cannery Sewer Fee Cut Sought." *San Jose News*, February 26.
———. 1976c. "Huge Sewer Fee Increase Approved for Canneries." *San Jose News*, March 3.
Gurtunya, Hulya Z. 1992. *An Analysis of the Transformation of a Social Problem: Occupational Health in Silicon Valley's Semiconductor Industry*. Unpublished Ph.D. dissertation, University of California, Santa Cruz.
Gutierrez, Ramon, and Richard J. Orsi (eds.). 1998. *Contested Eden: California before the Gold Rush*. Berkeley: University of California Press.
Gwynne, Robert N. 1999. "Globalization, Neoliberalism, and Economic Change in South America and Mexico." In *Latin America Transformed: Globalization and Modernity*. New York: Oxford University Press.
Ha, K. Oanh. 2000a. "Push Hard on Labor, Safety, Panel is Urged." *San Jose Mercury News*, March 24.
———. 2000b. "Piecework Lawsuit Settled Claim Over Man's Home Assembly of Electronics." *San Jose Mercury News*, November 14.
Hackel, Steven. 1998. "Land, Labor, and Production: The Colonial Economy of Spanish and Mexican California." In Ramon Gutierrez and James Orsi (eds.), *Contested Eden: California before the Gold Rush*, 111–146. Berkeley: University of California Press.
Halstead, Ted, and Clifford Cobb. 1996. "The Need for New Measurements of Progress." In Jerry Mander and Edward Goldsmith (eds.), *The Case against the Global Economy*, 197–206. San Francisco: Sierra Club Books.
Hammonds, Keith. 2000. "It's the Ideas, Stupid." *Dartmouth Alumni Magazine*, April, 16–17.
Hansen, Dirk. 1982. *The New Alchemists: Silicon Valley and the Microelectronics Revolution*. Boston: Little, Brown.
Hanson, Sam. 1969. "Civil Rights Group Eyes Union Tactics." *The Mercury*, March 1.
Harden, Blaine. 2001. "The Dirt in the New Machine." *New York Times Magazine*, August 12, 35–39.
Hardie, Crista. 1991. "EPA's Project Draws Heat." *Electronic News* 42, no. 2144 (November 25): 1.
Harper, Suzanne. 1986. "Hi Tech." *Occupational Health and Safety* (October).
Harris, Tom. 1982. "Sampling Error Delayed Discovery of Tainted Water." *San Jose Mercury*, February 3.
Hawes, Amanda. 1998. "Reflections of SCCOSH Founders." In *Silicon Dreams*. Santa Clara Center for Occupational Safety and Health's Workers' Memorial Day Book. May 3.
———. 2000. Various depositions from former employees of Silicon Valley firms. Law Offices of Alexander, Hawes, and Audet, San Jose, California.

Hayes, Dennis. 1989. *Behind the Silicon Curtain—The Seduction of Work in a Lonely Era*. Boston: South End Press.

Hayhurst, Chris. 1997. "Toxic Technology and the Silicon Valley." *E Magazine* 8, no. 3 (June): 19.

Hazardous Waste Contamination of Water Resources. 1985. Hearing before the Subcommittee on Investigations and Oversight of the Committee on Public Works and Transportation, U.S. House of Representatives. Washington, D.C.: Government Printing Office.

Health and Environmental Justice Project. 2001. "What Is Environmental Racism in Silicon Valley?" HEJ Project, San Jose, Calif., June. Silicon Valley Toxics Coalition, San Jose, California.

Heizer, Robert F., and A. E. Treganza. 1970. "Mines and Quarries of the Indians of California." In Robert F. Heizer and M. A. Whipple (eds.), *The California Indians: A Source Book*. Berkeley: University of California Press.

Henderson, Jeffrey. 1989. *The Globalisation of High Technology Production: Society, Space, and Semiconductors in the Restructuring of the Modern World*. London: Routledge.

Heritage Media Corporation. 1996. *Reflections of the Past: An Anthology of San Jose*. Encinatas, Calif.: Heritage Media Corporation.

Holliday, J. S. 1999. *Rush for Riches: Gold Fever and the Making of California*. Berkeley: Oakland Museum of California and University of California Press.

Holterman, Jack. 1970. "The Revolt of Estanislao." *The Indian Historian* 3: 44.

Hondagneu-Sotelo, Pierrette. 1994. *Gendered Transitions: Mexican Experiences of Immigration*. Berkeley: University of California Press

———. 2001. *Domestica: Immigrant Workers Cleaning and Caring in the Shadows of Affluence*. Berkeleya: University of California Press.

Hossfeld, Karen. 1988. *Division of Labor, Division of Lives: Immigrant Women Workers in Silicon Valley*. Unpublished dissertation, University of California, Santa Cruz.

———. 1990. "'Their Logic against Them': Contradictions in Sex, Race, and Class in Silicon Valley." In Kathryn Ward (ed.), *Women Workers and Global Restructuring*. Ithaca, N.Y.: ILR Press.

———. 1991. "Why Aren't High-Tech Workers Organized?" In *Common Interests: Women Organising in Global Electronics*. London: Working Women World Wide.

———. Forthcoming. *Small, Foreign, and Female*. Berkeley: University of California Press.

Huddle, Donald. 1985. *Illegal Immigration: Job Displacement and Social Costs*. Alexandria, Va.: American Immigration Control Foundation.

———. 1993. *The Costs of Immigration*. Washington, D.C.: Carrying Capacity Network.

———. 1994. *The Net Costs of Immigration to Texas*. Washington, D.C.: Carrying Capacity Network.

Hunter, Lori. 2000. "The Spatial Association between U.S. Immigrant Residential Concentration and Environmental Hazards." *International Migration Review* 34: 460–488.

Hurley, Andrew. 1995. *Environmental Inequalities: Class, Race and Industrial Pollution in Gary, Indiana, 1945-1980*. Chapel Hill: University of North Carolina Press.

Hurtado, Albert. 1992. "Sexuality in California's Franciscan Missions: Cultural Perceptions and Sad Realities." *California History* 71: 370–386.

Huws, Ursula. 1999. "Material World: The Myth of the 'Weightless Economy.'" *Socialist Register.* http://www.yorku.ca/socreg/.

Hyde, Charles Gilman, and George Leonard Sullivan. 1946. *Santa Clara County Sewage Disposal Survey. Report Upon the Collection, Treatment and Disposal of Sewage and Industrial Wastes of Santa Clara County California*. July 31. Board of Consulting Engineers. California Room, San Jose Public Library.

Industrial Survey Associates. 1950. *Santa Clara County: Its Prospects of Prosperity*.

Institute of Medicine. 1999a. *Toward Environmental Justice: Research, Education, and Health Policy Needs*. Washington, D.C.: National Academy Press.

Institute of Medicine. 1999b. *Addressing the Physician Shortage in Environmental Medicine*. Washington, D.C.: National Academy Press.

Jackson, Robert H. 1992. "The Dynamic of Indian Demographic Collapse in the San Francisco Bay Missions, Alta, California, 1776–1840." *The American Indian Quarterly* 16: 151–152.

———. 1994. "The Development of San Jose Mission, 1797–1840." In Lowell John Bean (ed.), *The Ohlone: Past and Present: Native Americans of the San Francisco Bay Region*, 229–247. Menlo Park, Calif.: Ballena Press.

Jacobson, Yvonne. 1984. *Passing Farms, Enduring Values: California's Santa Clara Valley*. Los Altos, Calif.: William Kaufman Inc.

Jacoby, Karl. 2001. *Crimes against Nature: Squatters, Poachers, Thieves, and the Hidden History of American Conservation*. Berkeley: University of California Press.

Jayadev, Raj. 2001a. "Winning and Losing Workplace Safety on Silicon Valley's Assembly Lines." *Pacific News Service*, January 18.

———. 2001b. *News Hour.* British Broadcasting Corporation. May 8.

Joachim, Leland. 1979. "Inflation, Job Losses Hit Minorities Hardest." *San Jose Mercury News*, January 29.

Johnson, John H. Jr., and Melvin Oliver. 1989. "Blacks and the Toxics Crisis." *The Western Journal of Black Studies* 13: 72–78.

Johnson, Missie. 1996. *Struggle & Strength: Tales from the Workers Stories Process*. May 1. San Jose, Calif.: SCCOSH.

Johnson, Steve. 1992. "Minorities Are the Majority in 137 Occupations in County." *San Jose Mercury News*, November 22.

Johnson, Susan Lee. 2000. *Roaring Camp: The Social World of the California Gold Rush*. New York: W. W. Norton.

Johnston, Moira. 1982. "High Tech, High Risk, and High Life in Silicon Valley." *National Geographic* 162: 459–476.

Jones, H. R. 1973. *Waste Disposal Control in the Fruit and Vegetable Industry*. Park Ridge, N.J.: Noyes Data Corporation.

Jung, Maureen. 1999. "Capitalism Comes to the Diggings." In James Rawls and Richard Orsi (eds.), *A Golden State: Mining and Economic Development in Gold Rush California*. Berkeley: University of California Press.

Kadetsky, Elizabeth. 1993. "High-Tech's Dirty Little Secret: Silicon Valley Sweatshops." *The Nation* 256: 517.

Karasek, Robert, and Tores Theorell. 1990. *Healthy Work: Stress, Productivity, and the Reconstruction of Working Life*. New York: Basic Books.

Katz, Naomi, and David Kemnitzer (eds.). 1983. "Fast Forward: The Internationalization of Silicon Valley." In June Nash and Maria Patricia Fernandez-Kelly (eds.), *Women, Men and the International Division of Labor*, 332–345. Albany: State University of New York Press.

———. 1984. "Women and Work in Silicon Valley: Options and Futures." In Karen Brodkin Sacks and Dorothy Remy (eds.), *My Troubles Are Going to Have Trouble with Me*. New Brunswick, N.J.: Rutgers University Press.

Kay, Jane. 1987. "Perils of Working in Silicon Valley." *San Francisco Examiner*, April 10.

Kazis, Richard, and Richard Grossman. 1982. *Fear at Work: Job Blackmail, Labor and the Environment*. New York: Pilgrim Press.

Keller, Don. 1976. "Poison-Dumping in County Called 'Serious.'" *San Jose Mercury News*, April 3.

Keller, John. 1981. *The Production Worker in Electronics: Industrialization and Labor Development in California's Santa Clara Valley*. Unpublished Ph.D. dissertation, University of Michigan.

———. 1983. "The Division of Labor in Electronics." In June Nash and Maria Patricia Fernandez Kelly (eds.), *Women, Men, and the International Division of Labor*. Albany: State University of New York Press.

Kelly Services. 1999. "The Free-Agent Workforce Has Grown to More than 24 Million Workers." March 15. http://www.kellyservices.com.

Kennedy, Danny. 2000. "Gold Mining: Environmental Destruction." Presentation at the University of California, Berkeley, October 4.

———. 2001. Presentation on Project Underground's Actions Concerning Oil and Mining Operations around the World. Panel on Combating Environmental Injustice. People's Summit on Globalization, University of Colorado at Boulder, March.

Kenney, Martin. 2000. *Understanding Silicon Valley*. Stanford, Calif.: Stanford University Press.

Kent, T. J. Jr. 1970. *Open Space for San Francisco Bay: Organizing to Guide Metropolitan Growth*. Berkeley: University of California, Institute of Government Studies.

Khor, Martin. 1996. "Global Economy and the Third World." In Jerry Mander and Edward Goldsmith (eds.), *The Case against the Global Economy*. San Francisco: Sierra Club Books.

Kilborn, Peter. 1987. "2 in Cabinet Fight Sale to Japanese." *New York Times*, March 12.

Kim, Sonja, and Helen Kim. 1996. "AIWA San Jose Members Follow Up on Their Successful Workplace Health and Safety Peer Training." *AIWA News* (November).

King, Deborah. 1988. "Multiple Jeopardy, Multiple Consciousness." *Signs* 14, no. 1: 42–72.

Klinger, Karen. 1980. "Santa Clara County Cancer Death Rate Jumps." *San Jose Mercury*, July 31.

———. 1982. "Investigators May Never Know Why Birth Defects Occurred, Expert Says." *San Jose Mercury News*, February 6.

Kovaleski, Serge. 1998. "High Technology's Top Banana? Costa Rica Lures Intel, Other Industry Giants." *The Washington Post*, March 11.

Krane, Jim. 2001. "IBM Chief to Receive British Knighthood." *Associated Press Wire*, June 19.

Krauss, Celene. 1993. "Blue-Collar Women and Toxic-Waste Protests." In Richard Hofrichter (ed.), *Toxic Struggles: The Theory and Practice of Environmental Justice,* chap. 10. Philadelphia: New Society Publishers.

Krech, Shepard. 1999. *The Ecological Indian: Myth and History*. New York: W. W. Norton & Co.

Kutzmann, David. 1985. "Officials Challenge Low-Risk Assumption." *San Jose Mercury News*, October 12.

Kwong, Peter. 1997. *Forbidden Workers: Illegal Chinese Immigrants and American Labor*. New York: The New Press.

LaDou, Joseph. 1984. "The Not-So-Clean Business of Making Chips." *Technology Review* 87, no. 4 (May/June).

———. 1985. "Health Issues in the Microelectronics Industry." *Occupational Medicine* 1: 1–12.

———. 1994. "Health Issues in the Global Semiconductor Industry." *Annals of the Academy of Medicine* [Singapore] 23, no. 5.

LaDuke, Winona. 1993. "A Society Based on Conquest Cannot Be Sustained." In Richard Hofrichter (ed.), *Toxic Struggles: The Theory and Practice of Environmental Justice,* 98–106. Philadelphia: New Society Publishers.

Lamborn, Pat. 1983. Interviews with several electronics workers. Santa Clara Center for Occupational Safety and Health. San Jose, California. November.

Lancet. 1973. "The Medical/Industrial Complex." 2: 1380–1381.

Landler, Mark. 2001. "Their Financial Crisis Past, Thais Remain Disillusioned." *New York Times*, June 26, C1.

Lapin, Lisa. 1989. "Landmark Toxics Law Up for OK in Valley." *San Jose Mercury News*, January 9.

Larimore, Jim. 1969. "La Raza Demands Minority Hiring Plan for Renewal Projects." *San Jose Mercury*, October 17.

Lasuen, Fermin F. 1965. *The Writings of Fermin Francisco de Lasuen*, Finbar Kenneally (ed.). Richmond, Va.: Academy of American Franciscan History.

Lavelle, Marianne, and Marcia Coyle. 1992. "Unequal Protection: The Racial Divide in Environmental Law." *National Law Journal* 15: S1–S12.

Lazaroff, Cat. 2001a. "Fish Can Be Hazardous to Your Health." *Environment News Service*, February 20.

———. 2001b. "Recycling Program Keeps Computers Out of Landfills." *Environment News Service*, May 23.

Lazarus, David. 2000. "Toxic Technology." *San Francisco Chronicle*, December 3.

Lee, Chong-Moon, William Miller, Marguerite Gong Hancock, and Henry Rowen. 2000. *The Silicon Valley Edge: A Habitat for Innovation and Entrepreneurship*. Stanford, Calif.: Stanford University Press.

Lee, Pam Tau. 1995. "Asian Workers in the U.S.—A Challenge for Labor." *Amerasia* 2.

Levander, Michelle. 1992. "Valley Lags Nation in Female Directors." *San Jose Mercury News*, July 2.

Leventhal, Alan, Les Field, Hank Alvarez, and Rosemary Cambra. 1994. "The Ohlone: Back from Extinction." In Lowell John Bean (ed.), *The Ohlone: Past and Present: Native Americans of the San Francisco Bay Region*, 297–336. Menlo Park, Calif.: Ballena Press.

Levy, Richard. 1978. "Costanoan." In Robert F. Heizer (ed.), *Handbook of North American Indians*, vVol. 8 (California), 485–495. Washington, D.C.: Smithsonian Institution.

Lewis, Sanford. 1997. "Feel-Good Notions, Corporate Power, and the Reinvention of Environmental Law." Working paper, Good Neighbor Project for Sustainable Industries.

Limbaugh, Ronald. 1999. "Making Old Tools Work Better: Pragmatic Adaptation and Innovation in Gold-Rush Technology." In James Rawls and Richard Orsi (eds.), *A Golden State: Mining and Economic Development in Gold Rush California*, 24–51. Berkeley: University of California Press.

Limerick, Patricia. 2000. *Something in the Soil: Legacies and Reckonings in the New West*. New York: W. W. Norton.

Lopez-Garza, Marta, and David R. Diaz (eds.). 2001. *Latino Immigrants in a Restructuring Economy: The Metamorphosis of Southern California*. Stanford, Calif.: Stanford University Press.

Lowen, Rebecca Sue. 1990. "'Exploiting a Wonderful Opportunity': Stanford University, Industry, and the Federal Government, 1937–1965." Ph.D. dissertation, Stanford University.

Lozano, Beverly Ann. 1985. "High Technology, Cottage Industry: A Study of Informal Work in the San Francisco Bay Area." Ph.D. dissertation, University of California, Berkeley.

Lukes, Timothy, and Gary Okihiro. 1985. *Japanese Legacy: Farming and Community Life in California's Santa Clara Valley*. Local History Studies 31. Cupertino, Calif.: California History Center.

MacLachlan, Malcolm. 1998. "Chip Industry Raises Environmental Concerns." *TechWeb*, December 1, http://www.techweb.com/wire/story/TWB19981201S0011.

Malone, Michael. 1985. *The Big Score: The Billion-Dollar Story of Silicon Valley*. Garden City, N.Y.: Doubleday.

Malone, Michael, and Susan Yoachum. 1980. "The Chemistry of Electronics." *San Jose Mercury News*, April 7.

Manan, Romulon. 2000. "Immigrant Workers and Temporary Labor." Presentation to the National Conference of Committees on Occupational Safety and Health, Boston, Massachusetts.

Mandich, Mitchell. 1975. *The Growth and Development of San Jose*. Master's thesis, San Jose State University.

Mann, Ralph. 1982. *After the Gold Rush: Society in Grass Valley and Nevada City, California, 1849–1870*. Stanford, Calif.: Stanford University Press.

Marbury, Hugh. 1995. "Hazardous Waste Exportation: The Global Manifestation of Environmental Racism." *Vanderbilt Journal of Transnational Law* 28.

Markoff, John. 1980. "California's Space-Age Sweatshops." *Los Angeles Times*, October 28.

———. 1992a. "Miscarriages Tied to Chip Factories." *New York Times*, October 12.

———. 1992b. "Miscarriage Risk for Chip Workers." *New York Times*, December 3.

———. 1999. "Boom Mind-Set Deeply Etched in Chip Capital." *New York Times*, June 16.

Marshall, Carolyn. 1987. "Fetal Protection Policies: An Excuse for Workplace Hazard." *The Nation*, April 25, 532–536.

Marshall, Jonathan. 1990. "Immigration Bill Triples Openings for Skilled Workers." *San Francisco Chronicle*, November 24.

Matthews, Glenna Christine. 1976. *A California Middletown: The Social History of San Jose in the Depression*. Ph.D. dissertation, Stanford University.

Matthews, John A., and Dong-Sung Cho. 2000. *Tiger Technology: The Creation of a Semiconductor Industry in East Asia.* Cambridge: Cambridge University Press.

Mazurek, Jan. 1999. *Making Microchips: Policy, Globalization, and Economic Restructuring in the Semiconductor Industry.* Cambridge, Mass.: MIT Press.

McCarthy, Francis Florence. 1958. *The History of Mission San Jose California, 1797–1835.* Fresno, Calif.: Academy Library Guild. Reprinted 1977.

McWilliams, Carey. 1935 [1949]. *California, the Great Exception.* New York: Current Books.

———. 1949. *North from Mexico: The Spanish-Speaking People of the United States.* New York: Greenwood Press.

———. 1969. *Factories in the Field: The Story of Migratory Farm Labor in California.* Hamden, Conn.: Archon Books.

Medoff, Peter, and Holly Sklar. 1994. *Streets of Hope: The Fall and Rise of an Urban Neighborhood.* Boston: South End Press.

Melosi, Martin. 1981. *Garbage in the Cities: Refuse, Reform, and the Environment, 1880–1980.* College Station: Texas A&M University Press.

Mendoza, Jay. 2001. "Health and Environmental Justice Project Highlights. May." Memo sent to HEJ Project board members and friends. Silicon Valley Toxics Coalition, San Jose, California.

Min, Pyong Gap. 1996. "The Entrepreneurial Adaptation of Korean Immigrants." In Silvia Pedraza and Ruben G. Rumbaut (eds.), *Origins and Destinies: Immigration, Race, and Ethnicity in America,* 302–314. New York: Wadsworth.

Molina, Alfonso Hernan. 1989. *The Social Basis of the Microelectronics Revolution.* Edinburgh: Edinburgh University Press.

Montague, Peter. 1998. "The Precautionary Principle." *Silicon Valley Toxics Action* (spring).

Morris, Jim. 1998. "High-Tech Industry Challenges Safety Rules." *Houston Chronicle,* September 27.

Mpanya, Mutombo. 1992. "The Dumping of Toxic Waste in African Countries: A Case of Poverty and Racism." In Bunyan Bryant and Paul Mohai (eds.), *Race and the Incidence of Environmental Hazards: A Time for Discourse,* chap. 15. Boulder, Colo.: Westview Press.

Mutz, Kathryn, Gary Bryner, and Douglas Kenney (eds.). 2002. *Justice and Natural Resources: Concepts, Strategies, and Applications.* Washington, D.C.: Island Press.

Nash, June, and Maria Patricia Fernández-Kelly. 1983. *Women, Men, and the International Division of Labor.* Albany: State University of New York Press.

National Industrial Pollution Control Council. 1971. *Pollution Problems in Selected Food Industries.* Washington, D.C.: NIPCC.

Navarro, Vincente. 1982. "The Labor Process and Health: A Historical Materialist Interpretation." *International Journal of Health Services* 12: 5–29.

Nelkin, Dorothy, and Michael Brown. 1984. *Workers at Risk: Voices from the Workplace*. Chicago: University of Chicago Press.

Northern California Labor. 1984. "SCC Toxic Coalition Keeps Pressure on EPA." San Francisco Edition. Vol. 37, no. 6, October 12.Nunis, Doyce B. Jr. 1964. "Letter from Dry Diggings, Alta, California, January 11, 1850." In *Letters of a Young Miner, 1849–1852,* 23. San Francisco: John Howell Books.

Nyden, Phil, Anne Figert, Mark Shibley, and Darryl Burrows. 1997. *Building Community: Social Science in Action.* Thousand Oaks, Calif.: Pine Forge Press.

O'Brien, Mary. 2000. *Making Better Environmental Decisions.* Cambridge, Mass.: MIT Press.

Organization for Economic Cooperation and Development (OECD). 1968. *Electronic Components: Gaps in Technology.* Paris: OECD.

Omi, Michael, and Howard Winant. 1994. *Racial Formation in the United States: From the 1960s to the 1990s.* 2d edition. New York: Routledge.

Overland Monthly. 1869. *Footprints of Early California Discoverers*, 261. Found in Bancroft Library, University of California Berkeley (cf. Bean 1994).

Page, Jake. 2000. "Making the Chips that Run the World." *Smithsonian* (January): 40.

Park, Edward. 1992. *Asian Americans in Silicon Valley: Race and Ethnicity in the Postindustrial Economy*. Unpublished Ph.D. thesis, University of California at Berkeley.

Park, Edward, and John Park. 1999. "A New American Dilemma?: Asian Americans and Latinos in Race Theorizing." *Journal of Asian American Studies* 2: 289–309.

Park, Lisa Sun-Hee, and David N. Pellow. 1996. "Washing Dirty Laundry: Organic-Activist Research in Two Social Movement Organizations." *Sociological Imagination* 33: 138–153.

Park, Lisa, and Grace Yoo. 2001. *The Impact of Welfare and Immigration Reforms on Immigrant Women's Access to Prenatal Health Care in California.* Report to the California Program on Access to Care, Berkeley, California.

Parker, Robert. 1994. *Flesh Peddlers and Warm Bodies.* New Brunswick, N.J.: Rutgers University Press.

Parkman, E. Breck. 1994. "The Bedrock Milling Station." In Lowell John Bean (ed.), *The Ohlone: Past and Present: Native Americans of the San Francisco Bay Region,* 43–63. Menlo Park, Calif.: Ballena Press.

Pasguini, D., and L. Laird. 1982. *Health Assessment of the Electronic Component Manufacturing Industry* (draft). Research Triangle Park, N.C.: Research Triangle Institute.

Passel, Jeffrey S. 1994. "Immigrants and Taxes: A Reappraisal of Huddle's 'The Cost of Immigrants.'" Washington, D.C.: The Urban Institute.

Passel, Jeffrey S., and Rebecca Clark. 1996. "Taxes Paid by Immigrants in Illinois." Technical Paper Produced for the Illinois Immigrant Policy Project. Washington, D.C.: Urban Institute.

Paules, Greta. 1993. *Dishing It Out: Power and Resistance among Waitresses in a New Jersey Restaurant*. Philadelphia: Temple University Press.

Pellow, David N. 1999a. "Silicon Valley Workplace Hazards: Centering the Workplace in Environmental Justice Research." Paper presented at the Annual Meetings of the American Sociological Association, Chicago, Illinois, August.

———. 1999b. "Negotiation and Confrontation: Environmental Policy-Making through Consensus." *Society and Natural Resources* 12: 189–203.

———. 1999c. "Framing Emerging Environmental Movement Tactics: Mobilizing Consensus, De-mobilizing Conflict." *Sociological Forum* 14: 659–683.

———. 2001. "Environmental Justice and the Political Process: Movements, Corporations, and the State." *The Sociological Quarterly* 42: 47–67.

———. 2002. *Garbage Wars: The Struggle for Environmental Justice in Chicago*. Cambridge, Mass.: MIT Press.

Pellow, David N., and Lisa Sun-Hee Park. 2000. "The Hazards of Work: Environmental Racism at the Point of Production." Paper presented at the Annual Meetings of the American Sociological Association, Washington, D.C., August.

Pena, Devon. 1997. *The Terror of the Machine: Technology, Work, Gender, and Ecology on the U.S.–Mexico Border*. Austin, Texas: Center for Mexican American Studies.

Perfecto, Ivette, and Baldemar Velasquez. 1992. "Farm Workers: Among the Least Protected." *EPA Journal* (March/April): 13–14.

Peterson, Paul E. 1981. *City Limits*. Chicago: University of Chicago Press.

Pisani, Donald J. 1999. "'I Am Resolved Not to Interfere, but Permit All to Work Freely': The Gold Rush and American Resource Law." In James Rawls and Richard Orsi (eds.), *A Golden State: Mining and Economic Development in Gold Rush California*, 123–148. Berkeley: University of California Press.

Pitt, Leonard. 1966. *The Decline of the Californios: A Social History of the Spanish-Speaking Californians, 1846–1890*. Berkeley: University of California Press.

Poling, Samantha. 1999. "Women Workers Not Told of Miscarriage Risk." *London*, February 17.

Pollack, Andrew. 1992. "It's Asians' Turn in Silicon Valley." *New York Times*, January 14.

Portes, Alejandro (ed.). 1995. *The Economic Sociology of Immigration: Essays on Networks, Ethnicity, and Entrepreneurship*. New York: Russell Sage Foundation.

Portes, Alejandro, Manuel Castells, and Lauren Benton (eds.). 1989. *The Informal Economy: Studies in Advanced and Less Developed Countries.* Baltimore, Md.: Johns Hopkins University Press.

Preston, William. 1998. "Serpent in the Garden: Environmental Change in Colonial California." In Ramon Gutierrez and Richard J. Orsi (eds.), *Contested Eden: California before the Gold Rush,* 260–298. Berkeley: University of California Press.

Prokopio, Mr. 1996. *Struggle & Strength: Tales from the Workers Stories Process.* San Jose, Calif.: SCCOSH.

Randol, James Butterworth. 1892. "Quicksilver." In David T. Day (ed.), *Report on Mineral Industries in the United States at the Eleventh Census: 1890,* 179–242. Washington, D.C.: Government Printing Office.

Rawls, James, and Richard Orsi (eds.). 1999. *A Golden State: Mining and Economic Development in Gold Rush California.* Berkeley: University of California Press.

Reisner, Marc. 1986. *Cadillac Desert: The American West and Its Disappearing Water.* New York: Penguin Books.

Richards, Bill. 1998. "Computer-Chip Plants Aren't as Safe and Clean as Billed, Some Say." *The Wall Street Journal,* October 5.

Rios, Delia. 1987. "Vietnamese Fear Cultural Backlash." *San Jose Mercury News,* July 26.

Rios, R., G. V. Poje, and R. Detels. 1993. "Susceptibility to Environmental Pollutants Among Minorities." *Toxicology and Industrial Health* 9 (September): 797–820.

Roberts, Bryan. 1995. "Socially Expected Durations and the Economic Adjustment of Immigrants." In Alejandro Portes (ed.), *The Economic Sociology of Immigration: Essays on Networks, Ethnicity, and Entrepreneurship,* 42–86. New York: Russell Sage Foundation.

Roberts, J. Timmons. 1993. "Psychosocial Effects of Workplace Hazardous Exposures." *Social Problems* 40: 74–89.

———. 1996. "Negotiating Both Sides of the Plant Gate." Paper presented at the Annual Meetings of the American Sociological Association, New York, August.

Roberts, J. Timmons, and Melissa Toffolon-Weiss. 2001. *Chronicles from the Environmental Justice Frontline.* New York: Cambridge University Press.

Robinson, Bert. 1986. "San Jose Council Votes against Toxics Initiative." *San Jose Mercury News,* October 1.

Robinson, J. Gregg, and Judith McIlwee. 1989. "Obstacles to Unionization in High-Tech Industries." *Work and Occupations* 16: 115–136.

Robinson, James. 1991. *Toil and Toxics.* Berkeley: University of California Press.

Roediger, David R. 1999. *The Wages of Whiteness: Race and the Making of the American Working Class.* Revised edition. New York: Verso.

Rogers, Everett, and Judith K. Larsen. 1984. *Silicon Valley Fever: The Growth of High-Technology Culture*. New York: Basic Books.

Rohrbough, Malcolm J. 1997. *Days of Gold: The California Gold Rush and the American Nation*. Berkeley: University of California Press.

Rollins, Judith. 1985. *Between Women: Domestics and Their Employers*. Philadelphia: Temple University Press.

Roske, Ralph J. 1968. *Everyman's Eden: A History of California*. New York: Macmillan.

Rossi, Mark Antony. 2000. Review of Jeff Kelly (ed.), *Best of Temp Slave*. Madison, Wis.: Garret County Press.

Rugemer, Werner. 1987. "The Social, Human and Structural Costs of High Technology: The Case of Silicon Valley." *Nature, Society, and Thought* 1: 149–160.

Rumbaut, Ruben. 1996. "Origins and Destinies." In Silvia Pedraza and Ruben G. Rumbaut (eds.), *Origins and Destinies: Immigration, Race, and Ethnicity in America,* 21–42. New York: Wadsworth.

Sachs, Aaron. 1999. "Virtual Ecology: A Brief Environmental History of Silicon Valley." *World Watch* (January/February): 12–21.

Sacks, Karen Brodkin, and Dorothy Remy. 1984. *My Troubles Are Going to Have Troubles with Me*. New Brunswick, N.J.: Rutgers University Press.

Salazar, Chito. n.d. *Global Product Chains: Northern Consumers, Southern Producers, and Sustainability*. Draft report prepared for United Nations Environment Programme. http://iisd.ca/susprod/virtualpolicy2.htm.

Salcedo, Victor. 2000. "Once a Temp, Always a Temp." *YO! Youth Outlook Magazine* 10, no. 6 (September).

Salzinger, Leslie. 1997. "From High Heels to Swathed Bodies: Gendered Meanings under Production in Mexico's Export-Processing Industry." *Feminist Studies* 23: 549–574.

Sancho, Raquel. 1999. "'We Like Them Small, Foreign, and Female': Filipino Workers Fight to Survive Philips-Taiwan." *Worker's Memorial Day Commemorative Program Book*. Santa Clara Center for Occupational Safety and Health. San Jose, California. May.

San Jose Board of Trade. 1887. *Santa Clara County, California*. San Francisco: Bancroft.

San Jose Planning Commission. 1961. *Design for Tomorrow*. May.

Sanchez, George. 1997. "Face the Nation: Race, Immigration, and the Rise of Nativism in Late Twentieth Century America." *International Migration Review* 31: 1009–1030.

Sandmeyer, Elmer. 1972. "The Bases of Anti-Chinese Sentiment." In Roger Daniels and Spencer C. Olin Jr. (eds.), *Racism in California*. New York: Macmillan.

Sanger, David. 1987. "Japanese Purchase of Chip Maker Canceled after Objections in U.S." *New York Times*, March 17.

Santa Clara Center for Occupational Safety and Health (SCCOSH). 1982. "Silicon Valley's 'Love Canal.'" *SCCOSH News* 3, no. 1 (February-March).

———. 1995. *Chemical Exposure Guidelines.* Version 9. October. San Jose, Calif.: SCCOSH.

———. 2000a. "If You Are a Temporary Worker." Printed fact sheet. San Jose, Calif.: SCCOSH.

———. 2000b. "Testimony before the Senate Industrial Relations Committee." March 23. SCCOSH, San Jose, California.

———. 2000c. "Electronics Assemblers: Protect Your Family from Lead." San Jose, Calif.: SCCOSH.

Santa Clara Center for Occupational Safety and Health and the South Bay Central Labor Council. 2000. "OSHA's Turnaround." *The Washington Post*, January 18.

Santa Clara County Planning Committee. 1973. *An Urban Development Open Space Plan for Santa Clara County, 1973–1978.* San Jose, Calif.: SCCPC. California Room, San Jose Public Library.

Sassen, Saskia. 1998. *The Mobility of Labor and Capital: A Study in International Investment and Labor Flow.* New York: Cambridge University Press.

Saxenian, AnnaLee. 1980. "Silicon Chips and Spatial Structure: The Industrial Basis of Urbanization in Santa Clara County, California." Master's thesis, University of California at Berkeley.

———. 1981. "Silicon Chips and Spatial Structure: The Industrial Basis of Urbanization in Santa Clara County, California." Working Paper 345, University of California Institute of Urban and Regional Development, Berkeley.

———. 1984a. "The Urban Contradictions of Silicon Valley: Regional Growth and the Restructuring of the Semiconductor Industry." In Larry Sawer and William K. Tabb (eds.), *Sunbelt/Snowbelt: Urban Development and Regional Restructuring*, 163–200. New York: Oxford University Press.

———. 1985a. "The Genesis of Silicon Valley." In Peter Hall and Ann Markusen (eds.), *Silicon Landscapes.* Boston: Allen and Unwin.

———. 1985b. "Silicon Valley and Route 128: Regional Prototypes or Historic Exceptions?" In Manuel Castells (ed.), *High Technology, Space, and Society*, 81–105. Beverly Hills, Calif.: Sage Publications.

———. 1994. *Regional Advantage: Culture and Competition in Silicon Valley and Route 128.* Cambridge: Harvard University Press.

———. 2000. "Networks of Immigrant Entrepreneurs." In Chong-Moon Lee, William Miller, Marguerite Gong Hancock, and Henry Rowen (eds.), *The Silicon Valley Edge: A Habitat for Innovation and Entrepreneurship*, 248–275. Stanford, Calif.: Stanford University Press.

Sayers, Ann Marie. 1994. "*NOSO-N*— 'In Breath So It Is in Spirit': The Story of Indian Canyon." In Lowell John Bean (ed.), *The Ohlone: Past and Present:*

Native Americans of the San Francisco Bay Region, 337–356. Menlo Park, Calif.: Ballena Press.

"SCC Toxic Coalition Keeps Pressure on EPA." 1984. *Northern California Labor,* San Francisco edition, 37, no. 6 (October 12).

Schalit, Naomi. 1984. "S. J. May Challenge Polluters." *San Jose Mercury,* May 25.

Schenker, Marc. 1992. *Epidemiologic Study of Reproductive and Other Health Effects among Workers Employed in the Manufacture of Semiconductors.* Final Report to the Semiconductor Industry Association. University of California at Davis.

Schnaiberg, Allan. 1973. "Politics, Participation and Pollution: The 'Environmental Movement.'" In John Walton and Donald Carns (eds.), *Cities in Change: Studies on the Urban Condition.* Boston: Allyn & Bacon.

———. 1980. *The Environment: From Surplus to Scarcity.* New York: Oxford University Press.

———. 1994. "The Political Economy of Environmental Problems." *Advances in Human Ecology* 3: 23–64.

Schnaiberg, Allan, and Kenneth A. Gould. 1994. *Environment and Society: The Enduring Conflict.* New York: St. Martin's Press.

Scott, Mel. 1959. *The San Francisco Bay Area: A Metropolis in Perspective.* Berkeley: University of California Press.

Sellers, Christopher. 1997. *Hazards of the Job: From Industrial Disease to Environmental Health Science.* Chapel Hill: University of North Carolina Press.

Semiconductor Industry Association. 1985. *Health, Safety and Environment Issues in the Semiconductor Industry.* San Jose, Calif.: SIA.

Sevilla, Ramon C. 1992. *Employment Practices and Industrial Restructuring: A Case Study of the Semiconductor Industry in Silicon Valley, 1955–1991.* Unpublished Ph.D. dissertation, University of California at Los Angeles, Department of Urban Planning.

Sferios, Emanuel. 1998. "Population, Immigration, & the Environment." *Z Magazine* (June): 24–29.

Sheehan, Helen, and Richard Wedeen (eds.). 1992. *Toxic Circles: Environmental Hazards from the Workplace into the Community.* New Brunswick, N.J.: Rutgers University Press.

Shoup, Laurence, and Randall Milliken. 1999. *Inigo of Rancho Posolmi: The Life and Times of a Mission Indian.* Ballena Press Anthropological Papers 47. Novato, Calif.: Ballena Press.

Siegel, Lenny. 1984. "High-Tech Pollution." *Sierra* 69 (November/December): 58–64.

———. 1994. *The Silicon Valley Experience: Why Labor Law Must Be Brought into the Twenty-First Century.* Testimony Before the U.S. Commission on the

Future of Worker-Management Relations. Mountain View, Calif.: Pacific Studies Center.

Siegel, Lenny, and Herb Borock. 1982. *Background Report on Silicon Valley*. Prepared for the U.S. Commission on Civil Rights. Mountain View, Calif.: Pacific Studies Center.

Siegel, Lenny, and John Markoff. 1985. *The High Cost of High Tech: The Dark Side of the Chip*. New York: Harper & Row.

Siegel, Lenny, Ted Smith, and Rand Wilson. 1990. "Sematech, Toxics, and U.S. Industrial Policy." *CPSR Newsletter* 8, no. 3 (summer): 9–10.

Silicon Valley Toxics Coalition, Campaign for Responsible Technology. 1996. "The Silicon Principles." San Jose, Calif.: Silicon Valley Toxics Coalition.

Silicon Valley Toxics Coalition. 1997. "What is SEMATECH?" http://www.svtc.org/sematech.htm.

———. 1998. "Health and Global Expansion of High-Tech." *Action Newsletter Archive* (spring).

———. 1999. "Toxic Timeline on Glycol Ethers." http://www.svtc.org, October 17.

———. 2001a. "Building an International Network." http://www.svtc.org.

———. 2001b. "International Campaign for Responsible Technology." http://www.svtc.org.

Simmons, William S. 1998. "Indian Peoples of California." In Ramon Gutierrez and Richard J. Orsi (eds.), *Contested Eden: California before the Gold Rush*. Berkeley: University of California Press.

Smith, Adam. 1910. *An Inquiry into the Nature and Causes of the Wealth of Nations*. London: J. M. Dent & Sons; New York: E. P. Dutton.

Smith, Michael, and Joe Feagin (eds.). 1987. *The Capitalist City*. Oxford: Basil Blackwell.

Smith, Rebecca. 1992a. "SEMATECH Funds Earmarked for Environmental Research." *San Jose Mercury News,* October 7.

———. 1992b. "Electronics Workers Ask for Help." *San Jose Mercury News,* October 20.

———. 1992c. "Chip Makers Promise Action on Toxics." *San Jose Mercury News,* December 4.

———. 1992d. "Groups Threaten Chip Firm Boycott." *San Jose Mercury News,* December 5.

Smith, Robert. 1998. "Closing the Door on Undocumented Workers." *NACLA Report on the Americas* 31, no. 4 (January/February): 6–9.

Smith, Ted. 1995. "Alternative Strategies." Memo to CSI Non-Governmental Organization Electronics group and allies. May 12.Silicon Valley Toxics Coalition, San Jose, California.

Smith, Ted, and Phil Woodward. 1992. *The Legacy of High-Tech Development: The Toxic Lifecycle of Computer Manufacturing*. San Jose, Calif.: Silicon Valley Toxics Coalition.

Snow, Robert T. 1983. "The New International Division of Labor and the U.S. Work Force: The Case of the Electronics Industry." In June Nash and Maria Patricia Fernandez-Kelly (eds.), *Women, Men, and the International Division of Labor*. Albany: State University of New York Press.

Sonnenfeld, David. 1993. "The Politics of Production and Production of Nature in Silicon Valley's Electronics Industry." Paper presented at the 88th Annual Meeting of the American Sociological Association, Miami, Florida, August.

Southwest Network for Environmental and Economic Justice and the Campaign for Responsible Technology. 1997. *Sacred Waters: Life-Blood of Mother Earth. Four Case Studies of High-Tech Water Resource Exploitation and Corporate Welfare in the Southwest*. Albuquerque, N.M.: SNEEJ.

Southwick, Karen. 1984. "High-Tech Industry Relies on Temporary Work Force." *San Jose Mercury News*, December 7.

St. Clair, David J. 1999. "The Gold Rush and the Beginnings of California Industry." In James Rawls and Richard Orsi (eds.), *A Golden State: Mining and Economic Development in Gold Rush California,* 185–208. Berkeley: University of California Press.

Stanley, Jerry. 2000. *Hurry Freedom: African Americans in Gold Rush California*. New York: Crown Publishers.

Stanley-Jones, Michael. 1998. "Silicon Valley Toxics Coalition Mapping Project Exposes Toxics in the 'hood." *Silicon Valley Toxics Coalition Action Newsletter Archive* (spring).

———. 2000. "Where Be-ist Mercury?" *Silicon Valley Toxics Action* (spring).

———. 2001. "Global RTK-Growing." *Silicon Valley Toxics Action* 19, no. 1 (spring): 3.

Stein, Walter J. 1973. *California and the Dust Bowl Migration*. Westport, Conn.: Greenwood Press.

Steinhart, Peter. 1985. "How Polluted Is My Valley?" *San Jose Mercury News*, October 20.

Stoecker, Randy. 1996. "Sociology and Social Action: An Introduction." *Sociological Imagination* 33: 3–17.

Stoecker, Randy, and Edna Bonacich. 1992. "Why Participatory Research." *American Sociologist* 23: 5–14.

Su, Julie. 1999. "The INS and the Criminalization of Immigrant Workers." In Joy James (ed.), *States of Confinement: Policing, Detention, and Prisons*. New York: St. Martin's Press.

Swanston, Samara F. 1999. "Environmental Justice and Environmental Quality Benefits: The Oldest, Most Pernicious Struggle and Hope for Burdened Communities." *Vermont Law Review* 23: 545.

Sweeney, Frank. 1997. "Report Criticizes Pollution Controls." *San Jose Mercury News,* November 21.

Szasz, Andrew, and Michael Meuser. 2000. "Unintended, Inexorable: The Production of Environmental Inequalities in Santa Clara County, California." *American Behavioral Scientist* 43: 602–632.

Taylor, Dorceta. 1993. "Minority Environmentalism in Britain." *Qualitative Sociology* 16: 263–295.

———. 1997. "American Environmentalism: The Role of Race, Class and Gender in Shaping Activism, 1820–1995." *Race, Gender and Class* 5: 16–62.

Teixeira, Lauren. 1997. *The Costanoan/Ohlone Indians of the San Francisco and Monterey Bay Area: A Research Guide.* Ballena Anthropological Papers 46. Menlo Park, Calif.: Ballena Press.

Thornton, Joe. 2000. *Pandora's Poison: Chlorine, Health, and a New Environmental Strategy.* Cambridge: MIT Press.

Thurm, Scott. 1990. "Firms Change to Help Save Ozone Layer." *San Jose Mercury News*, May 13.

———. 1992. "Toxics Lurk in Soil, Water." *San Jose Mercury News*, May 4.

———. 1993. "Firms Say It's Getting Harder to Cut Emissions." *San Jose Mercury News*, August 4.

Tilly, Chris. 1996. "The Good, The Bad, and the Ugly: Good and Bad Jobs in the United States at the Millennium." June. New York: Russell Sage Foundation.

Timberlake, Lloyd. 1987. *Only One Earth: Living for the Future.* London: BBC Books/Earthscan.

Torbert, Edward N. 1936. "The Specialized Commercial Agriculture of the Northern Santa Clara Valley." *Geographical Review* 26 (April): 247–263.

Townroe, P. 1969. "Locational Choice and the Individual Firm." *Regional Studies* 3: 15–24.

Trounstine, Philip, and Terry Christensen. 1982. *Movers and Shakers: The Study of Community Power.* New York: St. Martin's Press.

United Church of Christ. 1987. *Toxic Wastes and Race in the United States.* New York: United Church of Christ Commission for Racial Justice.

United States Bureau of the Census. 1973. *Census of the Population: 1970.* Volume 1, Characteristics of the Population, California. Washington, D.C.: U.S. Bureau of the Census.

United States Bureau of the Census. 1980. *Census of the Population: 1980.* Volume 1, Characteristics of the Population, California. Washington, D.C.: U.S. Bureau of the Census.

United States Bureau of the Census. 1990. *Census of the Population: 1990.* Volume 1, Characteristics of the Population, California. Washington, D.C.: U.S. Bureau of the Census.

United States Commission on Civil Rights. 1982. *Women and Minorities in High Technology: A Hearing before the USCR.* September 20–21. San Jose, Calif.: n.p.

United States Department of Health and Human Services. 1987. *NIOSH Alert. Request for Assistance in Reducing the Potential Risk of Developing Cancer from Exposure to Gallium Arsenide in the Microelectronics Industry.* Centers for Disease Control. Washington, D.C.: NIOSH.

———. 1989a. *Preliminary Health Assessment for Lorenz Barrel and Drum Company.* USDHHS, Public Health Service, Agency for Toxic Substances and Disease Registry. January 19.

———. 1989b. *Preliminary Health Assessment for Signetics Corporation.* USDHHS, Public Health Service, Agency for Toxic Substances and Disease Registry. January 19.

United States Department of Labor. 1971. *Work Injuries in the Canning and Preserving Industry.* Report No. 101. Washington, D.C.: Government Printing Office.

United States Environmental Protection Agency. 1992a. *Environmental Equity: Reducing Risk for All Communities.* Washington, D.C.: USEPA.

———. 1992b. "Special Issue: Environmental Protection: Has It Been Fair?" *EPA Journal* (March/April).

———. 1998. *United States–Mexico Border Environmental Indicators 1997.* Washington, D.C.: Government Printing Office.

United States General Accounting Office. 1983. *Siting of Hazardous Waste Landfills and Their Correlation with Racial and Economic Status of Surrounding Communities.* Washington, D.C.: Government Printing Office.

USA Today. 1998. "Dirty Secrets of the Chip-making Industry." January 12.

Vancouver, George. 1801. *A Voyage of Discovery to the North Pacific Ocean, and Round the World,* vol. 3. London: John Stockwell.

Viederman, Stephen. 1998. "Environmental Justice Investors to Ask Intel Shareholders for Support." Letter from the Jessie Smith Noyes Foundation, May 20. Jessie Smith Noyes Foundation, New York, N.Y.

Villones, Rebecca. 1989. "Women in the Silicon Valley." In Asian Women United of California (eds.), *Making Waves: An Anthology of Writings By and About Asian American Women,* 172–176. Boston: Beacon Press.

Wade, Richard, Michael Williams, Thomas Mitchell, Joel Wong, and Barbara Tuse. 1981. *Semiconductor Industry Study, 1981.* Sacramento: State of California, Department of Industrial Relations, Division of Occupational Safety and Health, Task Force on the Electronics Industry.

Wahlin, Britt. 2001. "Sweatshops, Silicon Valley Style." *Sojourner: The Women's Forum* (January).

Waller, Bill. 1980. "Tucson Hits the Industrial Big Time." *Tucson Weekly* News, February 20–26, 7.

Warren, Carol. 1991. *Schizophrenic Women in the 1950s.* Piscataway, N.J.: Rutgers University Press.

Wayne, Leslie. 2001. "Ex-Alcoa Boss May Become a Man of Steel." *New York Times,* July 17.
Webb, Edith B. 1952. *Indian Life at the Old Missions.* Lincoln: University of Nebraska Press.
Weinberg, Adam, David N. Pellow, and Allan Schnaiberg. 2000. *Urban Recycling and the Search for Sustainable Community Development.* Princeton, N.J.: Princeton University Press.
Welles, Edward. 1984. "Dirty Business in the Clean Room?" *San Jose Mercury News, West,* August 26.
Whyte, William. 1968. *The Last Landscape.* New York: Doubleday and Co.
Wilson, Tamar Diana. 1997. "Theoretical Approaches to Mexican Wage Labor Migration." In Darrell Y. Hamamoto and Rodolfo D. Torres, *New American Destinies: A Reader in Contemporary Asian and Latino Immigration,* 47–71. New York: Routledge.
Working Partnerships USA and the Economic Policy Institute. 1998. *Growing Together or Drifting Apart? Working Families and Business in the New Economy. A Status Report on Social and Economic Well-Being in Silicon Valley.* San Jose, Calif.: Working Partnerships USA and the EPI.
World Commission on Environment and Development. 1987. *Our Common Future.* New York: Oxford University Press.
Wright, Beverly, and Robert Bullard. 1993. "The Effects of Occupational Injury, Illness, and Disease on the Health Status of Black Americans." In Richard Hofrichter (ed.), *Toxic Struggles: The Theory and Practice of Environmental Justice,* 153–162. Philadelphia: New Society Publishers.
Yamane, Linda. 1994. "Costanoan/Ohlone." In Mary B. Davis (ed.), *Native America in the Twentieth Century: An Encyclopedia,* 143–144. Garland Reference Library of Social Science 452. New York: Garland Publishing.
Yoachum, Susan. 1982. "City's Reply to Contamination Questions Criticized." *San Jose Mercury,* March 31.
Zavella, Patricia. 1987. *Women's Work and Chicano Families: Cannery Workers of the Santa Clara Valley.* Ithaca, N.Y.: Cornell University Press.
Zinn, Maxine Baca. 1994. "Feminist Rethinking from Racial-Ethnic Families." In Maxine Baca Zinn and Bonnie Thornton Dill (eds.), *Women of Color in U.S. Society.* Philadelphia: Temple University Press.
Zlolniski, Christian. 1994. "The Informal Economy in an Advanced Industrialized Society: Mexican Immigrant Labor in Silicon Valley." *Yale Law Journal* 103: 2305–2335.

Index

About the Authors

DAVID NAGUIB PELLOW is Associate Professor in the Ethnic Studies Department at the University of California, San Diego. He is the author of *Urban Recycling and the Search for Sustainable Community* and *Garbage Wars: The Struggle for Environmental Justice in Chicago*. LISA SUN-HEE PARK is Assistant Professor of Ethnic Studies and Urban Studies and Planning at the University of California, San Diego.